Commercial Ship Surveying

Commercial Ship Surveying
On/Off-Hire Condition Surveys and Bunker Surveys

By

Harry Alexander Karanassos

AMSTERDAM • BOSTON • HEIDELBERG • LONDON
NEW YORK • OXFORD • PARIS • SAN DIEGO
SAN FRANCISCO • SINGAPORE • SYDNEY • TOKYO

Butterworth-Heinemann is an imprint of Elsevier

ELSEVIER

Butterworth Heinemann is an imprint of Elsevier
The Boulevard, Langford Lane, Kidlington, Oxford OX5 1GB, UK
225 Wyman Street, Waltham, MA 02451, USA

Notices
Knowledge and best practice in this field are constantly changing. As new research and experience broaden our understanding, changes in research methods, professional practices, or medical treatment may become necessary.

Practitioners and researchers must always rely on their own experience and knowledge in evaluating and using any information, methods, compounds, or experiments described herein. In using such information or methods they should be mindful of their own safety and the safety of others, including parties for whom they have a professional responsibility.

To the fullest extent of the law, neither the Publisher nor the authors, contributors, or editors, assume any liability for any injury and/or damage to persons or property as a matter of products liability, negligence or otherwise, or from any use or operation of any methods, products, instructions, or ideas contained in the material herein.

British Library Cataloguing in Publication Data
A catalogue record for this book is available from the British Library

Library of Congress Cataloging-in-Publication Data
A catalog record for this book is available from the Library of Congress

For information on all Butterworth Heinemann publications
visit our website at http://store.elsevier.com/

ISBN: 978-0-08-100303-9

Contents

List of Figures

List of Tables

Acknowledgments

I wish to thank the following, shown in alphabetical order, without whose help this book would not have been possible:

BIMCO

I am grateful to Bimco, who have very kindly agreed to the reproduction of:
Gentime, General Time Charter Party, Clause 5—On/Off-Hire Surveys.
Gentime, General Time Charter Party, Clause 6—Bunkers.

CARGOTECH-MACGRECOR

I am grateful to CargoTech-MacGregor, who have very kindly agreed to the reproduction of:
Numerous details of the construction of hatch cover parts and seal arrangements, as well as container securing equipment.

COOLERADO

I am grateful to Coolerado, who have very kindly agreed to the reproduction of:
Their psychrometric chart.

HAMBURG SUD

I am grateful to Hamburg Sud, who have very kindly agreed to the reproduction of:
Numerous high-resolution photos of their ships.

IACS

I am grateful to IACS, who have very kindly agreed to the reproduction of:
A number of very important diagrams of ships' parts.

IMO

I am grateful to IMO, who have very kindly agreed to the reproduction of:
The revised list of certificates and documents required to be carried on board ships.

LLOYD'S REGISTER FOUNDATION INFORMATION CENTRE LIBRARY

I am grateful to LR-FICL, who have very kindly agreed to the reproduction of:

A very important diagrammatic depiction of a bulk-carrier's cargo hold and the charts relating to the coatings condition in section 5.7.

PARKER-KITTIWAKE

I am grateful to Parker-Kittiwake, who have very kindly agreed to the reproduction of:

Photos covering their equipment regarding Fuel Oil density measuring apparatus and Fuel water content measuring apparatus.

ROLLS ROYCE

I am grateful to Rolls Royce, who have very kindly agreed to the reproduction of:

Numerous high-resolution photos covering their ship designs and marine equipment.

TEC CONTAINER

I am grateful to TEC Container, who have very kindly agreed to the reproduction of:

Container securing equipment.

Foreword

During a life of commercial activities, it is almost an unavoidable certainty that a ship will be involved in a number of legal disputes between her owners and her charterers. The disagreements which eventually lead to battles in court may arise as a result of the charterers refusing to make good damage sustained by the ship's hull and/or machinery, over the time period that the vessel was in their employment. Alternatively, the charterers may decline to arrange repairs of the damaged parts.

Damage to a ship's structure is not the only cause of disputes. It is possible that the charterers and the owners will become involved in a dispute concerning the quantities of fuel-oil and diesel-oil, which are claimed to have been consumed during the currency of a charter. The ship's owners will claim that the costs of these fuels are for the account of the charterers, whereas the latter will refuse settlement for their own reasons.

The time spent in courts, while the owners and the charterers try to prove who is right and who is responsible for the settling of the disputed amount, can be very expensive for both parties.

This book does not deal with ship design theories, or with the calculations relevant to Naval Architecture and Marine Engineering, that are related to the building of a vessel. However, it does cover marine surveys that are closely connected with the commercial operations of dry cargo ships.

It is hoped that the contents of this book will assist the surveyors who carry out on-hire and off-hire condition inspections to prevent the disputes which may arise between those who own ships and those who charter them, by writing factual, accurate, and—above all—well-defined reports on their findings.

SystÈme International D'Unites: [SI]

SI is the form of the metric system that is most widely used in both every day commerce and science. There are seven base units of measurement that have served for the derivation of 22 additional units as well as a set of prefixes that act as decimal multipliers.

Unit Name	Unit Symbol	Quantity Name	Dimension Symbol
Meter	m	Length	L
Kilogram	kg	Mass	M
Second	s	Time	T
Ampere	A	Electric current	I
Kelvin	K	Thermodynamic temperature	Θ
Mole	mol	Amount of substance	N
Candela	CD	Luminous intensity	J

SI Derived Units		
Base Quantity	Name	Symbol
Area	Square meter	m^2
Volume	Cubic meter	m^3
Speed, velocity	Meter per second	m/s
Acceleration	Meter per second squared	m/s^2
Wave number	Reciprocal meter	m^{-1}
Mass density	Kilogram per cubic meter	kg/m^3
Specific volume	Cubic meter per kilogram	m^3/kg
Current density	Ampere per square meter	A/m^2
Magnetic field strength	Ampere per meter	A/m
Amount-of-substance concentration	Mole per cubic meter	mol/m^3
Luminance	Candela per square meter	cd/m^2
Mass fraction	Kilogram per kilogram, which may be represented by the number 1	$kg/kg = 1$

SI Derived Units				
Derived Units	Name	Symbol	Other SI Units Expression	SI Base Units Expression
Plane angle	Radian	rad	1	m/m
Solid angle	Steradian	sr	1	m^2/m^2
Frequency	Hertz	Hz		s^{-1}
Force	Newton	N		$m\,kg\,s^{-2}$
Pressure, stress	Pascal	Pa	N/m^2	$m^{-1}\,kg\,s^{-2}$
Power, radiant flux	Watt	W	J/s	$m^2\,kg\,s^{-3}$

Electric charge, amount of electricity	Coulomb	C		s A
Electric potential difference, electromotive force	Volt	V	W/A	$m^2\,kg\,s^{-3}\,A^{-1}$
Capacitance	Farad	F	C/V	$m^{-2}\,kg^{-1}\,s^4\,A^2$
Electric resistance	Ohm	Ω	V/A	$m^2\,kg\,s^{-3}\,A^{-2}$
Electric conductance	Siemens	S	A/V	$m^{-2}\,kg^{-1}\,s^3\,A^2$

SI Derived units				
Derived Units	**Name**	**Symbol**	**Other SI Units Expression**	**SI Base Units Expression**
Magnetic flux	Weber	Wb	V s	$m^2\,kg\,s^{-2}\,A^{-1}$
Magnetic flux density	Tesla	T	Wb/m^2	$kg\,s^{-2}\,A^{-1}$
Inductance	Henry	H	Wb/A	$m^2\,kg\,s^{-2}\,A^{-2}$
Celsius temperature	Degree Celsius	°C		K
Luminous flux	Lumen	lm	cd sr[c]	Cd
Illuminance	Lux	lx	lm/m^2	$m^{-2}\,cd$
Activity referred to a radio nuclide	Becquerel	Bq		s^{-1}
Absorbed dose, specific energy (imparted), kerma	Gray	Gy	J/kg	$m^2\,s^{-2}$
Dose equivalent, ambient dose equivalent, directional dose equivalent, personal dose equivalent	Sievert	Sv	J/kg	$m^2\,s^{-2}$
Catalytic activity	Katal	kat		$s^{-1}\,mol$

SI Units Prefixes											
Prefix name		**Deca**	**Hecto**	**Kilo**	**Mega**	**Giga**	**Tera**	**Peta**	**Exa**	**Zetta**	**Yotta**
Prefix symbol		da	h	k	M	G	T	P	E	Z	Y
Factor	100	101	102	103	106	109	1012	1015	1018	1021	1024
Prefix name		Deci	Centi	Milli	Micro	Nano	Pico	Femto	Atto	Zepto	Yocto
Prefix symbol		d	c	m	μ	n	p	f	a	z	y
Factor	100	10−1	10−2	10−3	10−6	10−9	10−12	10−15	10−18	10−21	10−24

List of Abbreviations and Acronyms

The following abbreviations have been used throughout this book:

⊖	load line disk
⊗	Amidships mark
A	aft
AIS	automatic information system
ARPA	automatic radar plotting aid
A_w	water plane area
A_{ws}	wetted surface area
B	breadth
Blkhd	bulkhead
C_b	block coefficient
C_m	midship area coefficient
CM	cubic meter, m^3
COG	course over ground
C_p	prismatic coefficient
C_w	water plane area coefficient
D	depth
DB	double bottom
Deg	degree
DISPT	displacement, DISPLT, Δ
DWT	deadweight
ER	engine room
EXT	extreme
F	forward or for'd
Focs'le	forecastle
For'd	forward
GPS	global positioning system
H&M	hull & machinery
i.w.o.	in way of
I/D	propeller immersion as a percentage
IMO	International Maritime Organization
ISM	International Safety Management
KM, TKM	transverse metacentric height above the baseline
LBP	length between perpendiculars
LBP	length between perpendiculars
LCG	longitudinal center of gravity
LHS	left hand side
LKM	longitudinal metacentric height above baseline
LOA	length overall
LRFICL	Lloyd's Register Foundation Information Centre Library
LSA	life saving appliances

LT	long ton
LWL	load waterline
Mld	molded
MT	metric ton, tonne
MTC	moment to change trim one centimeter
OBO	ore, bulk, oil
OOW	officer on watch
P	port side
P&S	port and starboard
QMS	quality management system
RHS	right hand side
ROB	remaining on board
S	starboard-side or stbd-side
SAR	search and rescue
SF	stowage factor
SG	specific gravity
SOLAS	safety of life at sea
SSB	simply supported beam
SWL	safe working load (re. cargo gear)
TPC	tonnes per centimeter immersion
UDL	uniformly distributed load
UTC	universal coordinated time (also known as Z)
VCG	vertical center of gravity
WB	water ballast

Introduction

The foundation of ships' design is a combination of art, technology, and commercial purpose. The latest technologies concerning the equipment used for navigation, propulsion, and cargo handling form a strong influence on the final build. The multitude of cargoes carried by sea increases at great speed, and this necessitates the building of a more advanced ship design, more frequently. It is difficult to deal with the complexity of the foregoing, but the nature of man is such that he will always strive to find a solution to the problems at hand. A solution comes along sooner rather than later, and this triggers a demand for an even better product. This necessitates that the process starts all over again, but we must not forget that a successful progress brings greater rewards. Every commercial vessel is the servant of the most international of all industries, shipping. This industry was born in the depths of the most forgotten antiquity, and its purpose always was to convey cargoes and passengers. This tradition continues today, and this is where the ship's commercial life belongs.

It is the usual practice for a firm of shipowners to decide on the technical specifications of a new building in accordance with market requirements. Once a preliminary set of data is at hand, the owners' technical consultants investigate worldwide to find shipyards that can offer the closest to these requirements, at competitive prices. The new building must adhere to (1) the rules of the chosen classification society, (2) the guidelines of the flag state, and (3) the international maritime organization (IMO). Class surveyors will attend the new building at regular intervals to make sure that the methods and the materials used comply with the requirements of (1), (2), and (3). In due course, the vessel will commence her trading life, and she will be subject to periodical surveys/inspections. These surveys ensure that she maintains all the necessary standards and that she continues to satisfy the rules under which she is supposed to operate.

There are numerous different types of ships operating today in the various markets worldwide. One comes across bulk carriers, general cargo ships, high-speed craft, mobile offshore drilling units, passenger ships, oil tankers, container ships, Ro-Ro vessels, to name but a few. All those who design and build ships have a primary responsibility, as well as a concern, about the safe conveyance of crew, passengers, and cargo. The ship's officers and crew must put in place many safeguards so that the ship's sea-passage from the port of loading to the port of discharge is as safe as possible. Other aspects, equally as important, when designing the ship (e.g., ship's hull form, her engine-room outfit, her construction, and her equipment onboard) must be environmentally friendly. The designer must also ensure that there is sufficient protection of the environment from dangerous and noxious substances that a ship carries. The discharge of ballast and sediments is another area that needs careful consideration to prove that they do not create adverse effects in our world. Once designers, ship's officers, and engineers have solved all these technical adversities, the ship is ready to commence her commercial life.

The shipowner (for easy reference "Owner") makes his ship available to another party (for easy reference "Charterer"). In order to protect their interests, the Owner and the Charterer enter into a written agreement. This document defines the areas of their responsibilities, while it describes how the Charterer pays the Owner for letting his ship to him. It also details when the former may suspend payment(s). It is usual practice that the Charterer and the Owner appoint a competent surveyor to inspect the ship, on delivery to the Charterer. This inspection serves to put on record the condition of areas that the Charterer may utilize, as well as to ascertain the quantity of fuel (usually Heavy Fuel Oil and Marine Diesel Oil) remaining onboard at that time. A calculation of the quantity of "HF" and "MD" is essential because the Charterer reimburses the Owner with the cost of any fuel used, while the vessel operates for his account. We term this work as "ON-HIRE" survey. A similar assessment takes place when the agreement between Owner and Charterer ends. We term this work as "OFF-HIRE" survey. In the event that the ship suffered damage additional to that existing at the time of the initial survey, the Charterer is responsible to either arrange all necessary repairs or (if repairs are deferred) to pay the Owner for the cost of such repairs.

This book describes the details relevant to "ON-HIRE" and "OFF-HIRE" surveys, including the way in which one calculates the quantities of fuel oil remaining onboard and intends to convey a practical approach to the work involved, not only to those who have chosen ship surveys as their profession but also to those who serve onboard ships. Hopefully, the book will also offer an easy style of presentation. No book of this nature can hope to cover every aspect of all of the related theory and practice; accordingly, the reader will find a section, at the end of the book, which provides a number of references in respect of the subjects covered.

THE SHIPPING INDUSTRY

GETTING STARTED: SHIP TYPES

The shipping industry is continually concerned in endeavoring to anticipate what will be the next big source of trade. The fluctuations of world trade dictate which ships are those that will carry the most-in-demand products. Merchants and shipowners who are capable of taking advantage of such circumstances will have a better chance to profit the most. As far as shipowners are concerned, one can easily conclude that having the right ship at the right place where she is most sought after will result in high earnings.

As a consequence of the above, shipowners and shipbuilders have worked closely together to design and build different types of ships, which can handle and accommodate the widely varying requirements of the world's cargo markets.

Vessels which carry cargo can be divided into a number of groups, each offering a wide versatility of usage. The following listing is a compilation of the most common types of ships. Some of these ships carry cargoes while others just carry passengers, yet there ships that carry both cargo and passengers. Others ships are built to serve specialized requirements, that is, harbor tugs, icebreakers, and fishing vessels, to name but a few.

We are going to look at dry cargo ship types in more detail in Section 1.2.

DRY CARGO SHIPS

General cargo ships
Container ships
Roll-on/roll-off ships, including Car [& Passenger] Ferries
Multipurpose ships
Refrigerated cargo ships (Reefers)

BULK CARGO SHIPS

Dry bulk carriers
OBO carriers

LIQUID CARGO SHIPS

Crude oil tanker
Chemical tanker
Product tanker
VLCCs

GAS CARGO SHIPS

LPG carriers
LNG carriers

SPECIAL SERVICES SHIPS

High-speed crew boats and launches
Coastal ships
Harbor motor tugs
Deep sea tugs
Anchor handling/tug/supply vessels
Ice breakers
Fishing vessels
Barges
Heavy lift and semi-submersible ships

NOVEL DESIGN SHIPS

Hydrofoil crafts
Hovercrafts
Catamaran type passenger/car carriers.

PASSENGER SHIPS

Ocean cruise ships
Protected coastal waters and river cruise ships

DESCRIPTION OF SHIP TYPES: MORE DETAILS

So as to present a complete picture, the previous listing includes ships other than those carrying dry cargoes. However, because this work is dedicated to dry cargo ships, from this point on, we shall look exclusively at only this type of vessels.

DRY CARGO SHIPS: GENERAL CARGO SHIPS

After the end of the Second World War, dry cargo ships formed the bulk of the world's total merchant fleet. Those ships were known as general cargo vessels, and they were mainly outfitted with derricks, which enabled them to load and discharge cargo where there were no port cranes.

Some of those ships were loading and/or discharging goods over a set number of ports, and they were said to operate a "liner" service, whereas others followed completely irregular routes, always chasing opportunities of high paying rates.

Subsequent versions of these ships were introduced which were outfitted with heavy lift derricks, in addition to lighter duty derricks. Tween decks provided a good arrangement for the carriage of general cargoes, with heavier consignments loaded in the lower holds. The hatch covers of these ships were provided with fittings so that they could carry containers. The derricks of these ships, in the late seventies, were replaced by cranes. This modification resulted in cargo gear capable of facilitating faster cargo operations (Figures 1 and 2).

FIGURE 1

General cargo ship with cranes retrofitted.

FIGURE 2

General cargo ship. A crewboat is loaded on hatch no. 2 as deck cargo.

Source: Author's own library.

DRY CARGO SHIPS: CONTAINER SHIPS

A ship appeared in the early 1950s that would revolutionize the way cargoes were carried. Every cargo onboard this type of ship was stuffed in 20- or 40-ft. steel containers. Both types of these boxes had a cross section of 8×8 ft. Over the years, the dimensions of these boxes have changed to accommodate different trades; however, these changes have not been drastic (Figure 3).

FIGURE 3

Dry cargo ships—container ship.

Source: HAMBURG SÜD.

FIGURE 4

Dry cargo ships—container ship.

Source: HAMBURG SÜD.

Today, all types of cargoes are carried in containers, in an extremely orderly and secure way (provided the appropriate care has been taken by those who place the goods in the boxes). Particularly interesting are containers that carry refrigerated cargoes, such as fruits and deep frozen meat and fish (Figure 4).

In addition, container framework has been modified so that pressurized gases can be carried in tanks fitted within a 20-ft. container base. Today's largest container vessels are designed to be capable of carrying on deck and under deck, a total of 18,000 units.

DRY CARGO SHIPS: RO-RO SHIPS

In the days when it was not possible to construct bridges so that transports (including trains) could cross from one side to the other, ships with rails were employed. The train would roll on, and the ship and its cargo would then travel to the other side, where, at a convenient location, the train would roll off the ship and continue its journey.

That is the principle with which ferries are closely associated. One can say that "ship-landing-craft" was—perhaps—an early RO-RO design. In the middle of the nineteenth century, it became evident that carrying cars as well as their drivers/passengers was becoming not only a pressing requirement but also a profitable mode of sea transport.

There are RO-RO vessels that operate on transcontinental routes, from the port nearest to a car-building facility to a port where these cars would be placed for sale in the local markets. General cargoes can be in boxes, pallets, or unpacked. Some of these units can carry grain in bulk. Bulk cargoes and gas cargoes are also carried on suitably modified units.

Other units capable of carrying refrigerated cargoes, under controlled atmospheres, have also become part of the payload of a modern container ship. The early types of container ships carried approximately 100 units and were outfitted with cranes which could facilitate the loading and/or unloading of units; however, as time passed and these ships underwent changes in line with worldwide requirements and demands, we reached a point where ports are designed to accommodate these ships and to provide as good a platform for cargo operations as possible (Figures 5 and 6).

There are also RO-RO vessels that can carry private cars and heavy transports to nearby continents. These ships are outfitted with restaurants and sleeping cabins for the drivers and passengers of cars carried onboard. RO-RO ships of this configuration are designed so that they can carry the maximum number of cars and passengers (Figure 7).

FIGURE 5

Dry cargo ships—Car carriers.

FIGURE 6

Dry cargo ships—RO-RO ship, including car & passenger ferries.

Source: Rolls Royce.

FIGURE 7

Dry cargo ships—RO-RO ship, including car & passenger ferries.

DRY CARGO SHIPS: MULTIPURPOSE SHIPS

These ships have been built for the purpose of being able to carry a diverse variety of goods at the same time. Steel profiles, forest products, bagged cargoes, building materials, and earth movers are a few of the applicable examples. The majority of these ships are provided with their own cargo gear so as to be able to load and discharge irrespectively of the availability of cargo handling equipment (or the lack of it) at the ports they visit (Figure 8).

The cargo gear may be such as to be capable of dealing with heavy lifts. Equally, tank tops may be strengthened for the carriage of heavy point loads. The ships have the capability of visiting ports or anchorages where other ships cannot easily operate.

FIGURE 8

Dry cargo ships—multipurpose ship.

DRY CARGO SHIPS: REFRIGERATED CARGO SHIPS (REEFERS)

The cargoes they carry is perishable, and accordingly, these ships are designed to be fast. Often they will have white-painted hulls so as to reflect the sun's heat, thus helping to keep the cargo cool. Reefer ships—simply put—are large refrigerators, with heavy insulation. Their cargo holds are each divided into a lower hold and between one and three tween decks. Locker spaces may also be provided.

All these separate cargo spaces can carry different types of cargo at different temperatures. Today, most cargoes are carried and handled in palletized form, and they are moved about on conveyors. Some cargo types (such as bananas) are often handled through doors in the ship's side, and their handling is by means of electric fork lift trucks.

It is not difficult to appreciate that the two most important factors as far as reefer ships are concerned are cleanliness and the maintenance of optimum temperatures (Figure 9).

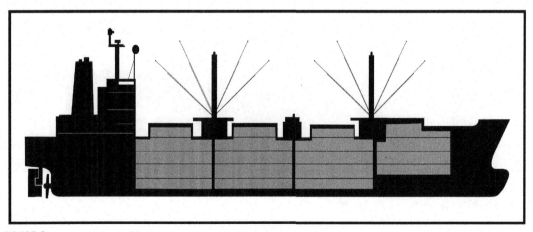

FIGURE 9

Typical configuration of a reefer ship, with E.R. and accommodation arranged aft.

Their cargo gear may comprise derricks or cranes, and they can carry reefer containers on deck and on their hatch covers.

Today, reefer ships face heavy competition from the large container carriers that can carry, in refrigerated containers, many times the tonnage of an average size conventional reefer vessel. Under the present circumstances, liner reefer companies will be the ones to survive if they diversify and pick their niches carefully

DRY CARGO SHIPS: DRY BULK CARRIERS

Bulk carriers made their appearance in the shipyards of the world in the 1950s. Without a doubt, this type of vessel has become the backbone of the shipping industry, and in the twenty-first century, it accounts for almost 33% of all commercial ships (Figure 10).

The design, construction, and operation of this type of ship has attracted considerable attention over the years, as it became evident that the speed of loading/discharging as well as the sequence of the holds (where cargoes were loaded/discharged) resulted in structural problems and even catastrophic failures.

As a result of these incidents, the calculations of certain strength members of the ship had to be reviewed, with additional material being introduced in the hull areas that required strength improvements.

The single-deck design format of the cargo holds is that of a totally unobstructed box. This enables the carriage of dry bulk cargoes, such as grain, iron, coal, and concentrates of iron, bauxite, and aluminum.

FIGURE 10

Dry cargo ships—geared dry bulk carrier.

Source: HAMBURG SÜD.

FIGURE 11

Dry cargo ships—geared dry bulk carrier.

Source: HAMBURG SÜD.

The cargo holds' assembly is a rectangular prism [or cuboid], with the accommodation, navigating bridge and engine room arranged at the after end and with the bow arranged at the forward end (Figure 11).

DRY CARGO SHIPS: OBO CARRIERS

An OBO carrier is a bulk carrier designed to carry dry bulk cargoes as well as liquid bulk cargoes. This design was based on the principle of the ship being able to be used as a tanker when the oil markets offered preferential rates and then to switch to dry bulk cargoes when the oil markets had subsided.

This way, such ships would enjoy the peaks of both markets. With early OBOs, the process of switching from liquid to dry cargoes proved to be both difficult as well as expensive.

Today, few operators are still involved with the operation of OBO ships; however, the popularity of these ships enjoyed in the 1970s is no longer evident among shipowners.

SPECIAL SERVICES SHIPS: HIGH SPEED CREW BOATS AND LAUNCHES

Crew boats are very important to the shipping industry as they provide the connection between a base onshore and offshore installations, such as drilling rigs, or designated anchorages which serve hundreds of ships at a time.

Crew boats are designed with a passenger cabin capable of accommodating between 20 and 30 passengers, with the navigating bridge arranged immediately forward of this cabin.

There is usually a large open deck aft capable of carrying some cargo (2-3 tonnes). They are fast craft with a crew of 4-5, including the master. Some ports allow these crew boats to operate 24 h a day, whereas others only allow operations during daylight hours. The usual range of this craft is 40-50 miles from their shore-based head office (Figure 12).

FIGURE 12

Special services ships—crewboats.

SPECIAL SERVICES SHIPS: COASTAL SHIPS

Coastal ships used to be relatively small vessels which operated between ports that were separated by small distances. Today, with modern machinery, equipment, and navigational aids, they may carry cargo and some passengers over greater distances than they used to. They are outfitted with some form of cargo gear (i.e., derricks or cranes) and they may have a tween deck (Figure 13).

Some of these small ships may also serve as bunker barges that supply larger ships or as fuel transporters to islands that are not too distant from the mainland.

FIGURE 13

Special services ships—coastal ships.

SPECIAL SERVICES SHIPS: HARBOR MOTOR TUGS

Harbor tugs are small vessels with big engines, which are essential in every port of the world because without them overseas carriers would not be able to dock or undock. The great majority of these vessels are powered by diesel engines; however, lately there are a number of these vessels outfitted with engines capable of using LNG as their main fuel. These ships are also provided with substantial gear for fire fighting (Figure 14).

FIGURE 14

Special services ships—harbor tugs.

Source: Rolls Royce.

SPECIAL SERVICES SHIPS: DEEP SEA TUGS (FIGURE 15)

Deep sea tugs are large vessels with powerful engines and with pumping arrangements that are capable of handling large quantities of water either for firefighting or for pumping out water that may be entering the towed vessel, through shell fractures or ruptured intakes/discharges. They carry large quantities of fuel so as to tow a vessel over long distances. Accommodation is available for the crew and for any riding crew that may be required by a disabled vessel. In view of the type of work these tugs undertake, their towing winches are of high capacity.

FIGURE 15

Special services ships—deep sea tugs.

SPECIAL SERVICES SHIPS: AHTS VESSELS (FIGURES 16 AND 17)

AHTS (Anchor Handling Tug Supply) vessels are designed to serve oil rigs. They can handle oil rig anchors and can also tow oil rigs to location and anchor them up. They transport supplies to and from offshore drilling rigs.

Occasionally, they can serve as Emergency Rescue and Recovery Vessels (ERRV). AHTS vessels have machinery which is specifically designed for anchor handling operations with power typically ranging from 5000 to 12,000 bhp, which can produce between 60 and 150 tonnes of bollard pull. They

FIGURE 16

Special services ships—AHTS vessel.

Source: Rolls Royce.

FIGURE 17

Special services ships—AHTS vessel.

Source: Rolls Royce.

also have arrangements for quick anchor release, which is operable from the bridge or other location which is normally manned and in direct communication with the bridge. The reference load used in the design and testing of the towing winch is twice the static bollard pull.

SPECIAL SERVICES SHIPS: ICEBREAKERS (FIGURE 18)

The design of an icebreaker is such that she can sail through thick ice at the head of a convoy of ordinary cargo ships, clearing a waterway for them. These ships are of a double hull construction that allows them to reach polar regions and carry out research. Their scantlings are increased when compared to other ordinary vessels, and the steel used in their construction is of a special grade that does not lose strength when exposed to very low temperatures.

Parts such as propellers and rudders are placed in positions that allow maximum protection by the hull. It is not uncommon for propellers to be fitted at the stern as well as at the bow, so as to afford enhanced operational maneuvers, over and above the requirements of a cargo ship in case ice becomes too restrictive to the ship's movement.

Bow thrusters are also provided. The bow of these ships is of special configuration so that it can ride on top of ice and break it under the ship's weight. The main engines and generators of these ships are of a power over and above that required for commercial vessels of the same size.

FIGURE 18

Special services ships—icebreakers.

SPECIAL SERVICES SHIPS: FISHING VESSELS

Fishing vessels today are like a sizeable processing factory. They have sophisticated equipment for tracking the best location for fishing, and they can stay at sea for long periods of time. They store their catch—once it is cut and suitably packed—in refrigerated cargo holds.

These holds are capable of maintaining the cargo in deep frozen condition, until it can be discharged at a suitable port from where it can be sent to the nearest fish markets. A fishing trawl-net is towed by the vessel and the catch is hauled up a stern ramp. The nets are weighted for bottom trawling or pelagic, which involves mid-water trawling.

All the fish that is caught is processed by a total of 30-40 crewmembers, upon coming onboard. The particular method of processing is carried out in accordance with the specifications of prospective buyers. Large fish factories can be as long as 140 m, and they can reefer-store up to 6000 tonnes of fish. The trawler below is a Rolls-Royce NVC 370 design and is 69 m long which will carry out traditional fishing operations for whitefish and shrimp.

The vessel is equipped with a hybrid propulsion system, based on the Rolls-Royce hybrid shaft generator system (HSG). It can be operated in diesel-electric or diesel-mechanical mode. Rolls-Royce will also supply electric permanent magnet-driven trawl winches to the vessel (Figures 19 and 20).

FIGURE 19

Special services ships—trawler vessel.

Source: Rolls Royce.

FIGURE 20

Special services ships—trawler vessel.

Source: Rolls Royce.

SPECIAL SERVICES SHIPS: BARGES (FIGURE 21)

A barge is a box-shaped vessel with tapered ends and a flat deck, which is outfitted with stanchions.

The stanchions support steel plating that protects the cargoes carried from sea-spray.

This boundary also serves to assist with the stowing and securing of the cargoes carried on the deck. The space between the deck and the bottom structure is subdivided longitudinally and transversely by watertight bulkheads. Ballasting these tanks trims the barge—as may be required—by load-on/off operations.

Barges are towed by a tug, as they have no propulsion engines of their own and whilst they usually carry cargoes of a low center of gravity, they possess sufficient stability to carry equipment of a considerable height (e.g. Container gantry cranes, or mobile grab cranes), weather permitting.

FIGURE 21

Barge maneuvered by tug.

SPECIAL SERVICES SHIPS: HEAVY LIFT AND SEMISUBMERCIBLE SHIPS

Heavy-lift ships are—thanks to their structural design and building—vessels capable of transporting extremely heavy loads that cannot be transported by ordinary ships. There are—broadly speaking—two types of heavy lift ships.

The first type is a tween deck vessel with accommodation and machinery arranged either forward or aft so as to allow for the maximum deck area, where heavy loads may be placed, lashed, and secured. The scantlings of such ships are considerably greater than those applicable to ordinary cargo ships. They have a deck-arranged heavy cargo gear, typically of capacities well over 150 tonnes. These ships are well suited for transporting and loading/discharging large and heavy cargoes in ports where the necessary facilities for this type of cargo operations do not exist. The second type of heavy lift ships are those that are constructed with their bridge and accommodation well forward, whereas their funnels and other exhausts are arranged, symmetrically on port and starboard sides, at their stern.

This configuration allows for a very large deck area. These ships are capable of submerging their deck well below sea-level, by taking onboard large quantities of ballast, in a precisely controlled manner. In this fashion, their cargoes (often being incapacitated ships and drilling rigs) can be floated over the deck and held at this position, while ballast water is pumped out. Through this mode of operation, the base of the cargo will eventually come to rest on the deck of the semisubmercible, ready for securing. Both of these types of heavy lift vessels are capable of transporting their cargoes over great distances.

While on the subject of ships capable of heavy lifts, we should mention a third type of vessel which is outfitted with very large capacity cranes (SWL: 12,000 tonnes), servicing shipbuilding yards and offshore drilling rigs and other similar installations (Figure 22).

FIGURE 22

Heavy lift ships, double banked for "ship-to-ship" cargo transfer (RHS crane capacity: 700 tonnes).

SPECIAL SERVICES SHIPS: OCEAN CRUISE SHIPS

A cruise ship carries passengers who can enjoy an array of state of the art facilities, including theaters, cinemas, spas, to name but a few (Figures 23 and 24).

The ship may call at a number of port-of-calls, each offering a unique experience and entertainment to her passengers. Eventually, after having visited a number of destinations, the passengers return to their originating port. An alternative type of cruise involves sailing for 2-5 days at sea and then returning to the port of embarkation. Ocean liners involved in transoceanic voyages are built to higher standards than a cruise ship so that they can withstand adverse sea conditions that prevail in the Atlantic or the Pacific. Ocean liners are considerably larger than cruise ships, thus incapable of entering ports of shallow draft.

Today passenger ships have evolved from ocean liners to cruise ships that have passenger cabins arranged on the superstructure, with verandas and large windows. In addition, they have added amenities to cater for swimming and other water sports. The latest designs have transformed what used to be a passenger ship into a floating community enjoying the highest luxury, entertainment, and sport including exercise facilities.

FIGURE 23

Ocean liners/luxury cruise ships.

FIGURE 24

Ocean liners/luxury cruise ships.

Transatlantic cruising today is part of the tourism industry, which is enjoying a rapid growth.

In order to increase passenger-volume coupled with a more affordable cost, cruising has taken a side step toward river trips and coastal trips in protected waters (some across different countries). Such trips take place on smaller size boats that offer good quality accommodation, fine cuisine, usual amenities (i.e., TV, PC, library), and excellent service (Tables 1 and 2).

Table 1 Ships' Dimensions in Accordance with Type—All Ships

Ship Type	Gross Tons	Net Tons	LOA	LBP	Molded B	Molded D	Summer Draft	DWT
Bulk carrier	12,000		157.90	151.50	23.20	12.50	8.60	20,000
Bulk carrier			95.00	90.00	16.00		6.10	5000
Bulk carrier			107.00	102.00	18.20		7.30	8000
Bulk carrier HS			117.00	110.00	19.30		7.80	10,000
Bulk carrier HS			153.00	143.00	23.20		9.10	20,000
Bulk carrier HS			170.00	163.00	27.00		10.50	30,000
Bulk carrier HS			189.99	186.65	30.00	15.00	10.70	43,500
Bulk carrier HM			178.00	170.00	28.00		9.50	35,000
Bulk carrier HM	20,471	10,630	178.80	170.00	27.20	14.20	10.43	27,659
Bulk carrier HM			185.00	177.00	30.40		10.50	45,000
Bulk carrier HM			190.00	183.00	32.26		11.50	55,000
Bulk carrier HM	33,631	19,329	193.00	185.00	32.26	18.50	13.02	57,411
Bulk carrier PM	38,217		200.00	191.00	36.00	18.45	11.60	67,000
Bulk carrier PM			225.00	217.00	32.26		14.20	75,000
Bulk carrier PM	44,325	26,919	229.00	222.00	32.26	20.20	15.10	82,000
Bulk carrier PM	38,878	24,416	225.00	217.00	32.25	19.00	14.00	72,824
Bulk carrier post PM			239.99		38.00		14.48	98,681
Bulk carrier CS			225.00	217.00	37.00		13.10	80,000
Bulk carrier CS			235.00	226.00	43.00		13.60	100,000
Bulk carrier CS			270.00	260.00	44.00		17.30	150,000
Bulk carrier CS			289.00	279.00	45.00		18.40	175,000
Bulk carrier CS	92,278		292.00	282.00	45.00	24.70	18.20	180,000
Very large bulk carrier			300.00	289.00	50.00		18.70	205,000

Ship Type	Gross Tons	Net Tons	LOA	LBP	Molded B	Molded D	Summer Draft	DWT
Very large bulk carrier			320.00	308.00	53.00		18.60	230,000
Very large bulk carrier			332.00	320.00	58.00		22.80	320,000
Container ship	4235	2100	108.50	102.40	15.88	7.70	5.93	5522
Geared multi purpose container ship	10,352	5804	158.06	148.03	22.00	11.00	7.98	13,387
Geared multi purpose container ship	18,391	9229	184.90	176.00	27.60	14.70	10.59	25,561
Geared multi purpose containership			199.90	190.00	28.20	15.50	10.50	31,000
General cargo ship + containers	16,800	6900	168.14		25.20		10.65	21,150
Refrigerated vessel	11,382	6408	155.00		24.00		10.10	12,902
Coastal ship + containers	2056	1168	88.95		12.50		4.34	2964
Ferry			172.90	160.58	25.70	9.40	6.35	4500

S, Handysize; HM, Handymax; PM, Panamax; CS, Capesize; Molded B, molded beam; Molded D, molded depth.

Table 2 Categories of Bulk-Carriers in Accordance with Deadweight

Bulk Carrier Designation	Deadweight		
Seawaymax	10,000-to- 60,000		
Handysize	26,000-to- 40,000		
Handymax	40,000-to- 50,000		
Supramax	50,000-to- 60,000		
Panamax	60,000-to- 80,000		
Post Panamax	Upto 125,000		
Capesize	125,000-to- 220,000	Kamsarmax	175,000
		Newcastlemax	185,000
		Setouchmax	203,000
Very Large Ore Carrier	Larger than 220,000		

SHIP STABILITY, FREEBOARD, AND COMMON TERMS

SHIP STABILITY

One of the most important topics of study in Naval Architecture is stability. As an overview, we shall look into the following states of equilibrium.

On a tranquil day, three boats are afloat and motionless, in a secluded anchorage.

Boat "ALPHA1" is floating upright in an apparent state of equilibrium. Momentarily, an external force is applied to this boat. The equilibrium of the boat is temporarily upset with the boat taking a new posture. However, after a short period of time, she moves back to her original upright posture. The boat is said to be in *a state of stable equilibrium*.

Boat "ALPHA2" is floating upright in an apparent state of equilibrium. Momentarily, an external force is applied to this boat. The equilibrium of the boat is temporarily upset with the boat taking a new posture, where she comes to rest. The boat is said to be in *a state of neutral stability*.

Boat "ALPHA3" is floating upright in an apparent state of equilibrium. Momentarily, an external force is applied to this boat. The equilibrium of the boat is upset to the extent that she continues to move away from her initial position. The boat is said to be in *a state of negative stability*.

We mentioned above that the boats are originally motionless. This is an important consideration because when ships are under way there may be a number of external forces that can affect their buoyancy.

For a boat to remain in a state of equilibrium, it is necessary that the following conditions are satisfied (Figure 25).

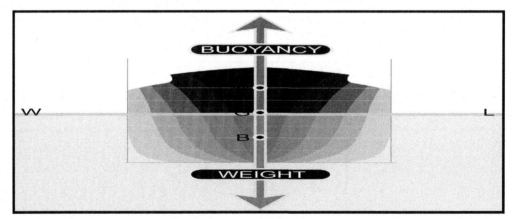

FIGURE 25

Equilibrium for a ship floating in still water.

Commercial Ship Surveying

The forces of weight and buoyancy must be equal. They must lie on the same vertical line.

A Naval Architect spends quite a proportion of his work-time establishing ship areas, their moments and their second moments. One of the common methods employed to calculate an area bound by a curve is by using *Simpson's First Rule*. The provision for using this rule is that the area under a curve is enclosed by an odd number of equispaced ordinates.

The area of KLNM is given by the formula:

$$\frac{h}{3} * (y1 + 4y2 + y3)$$

where *h* is the common space between the ordinates.

The method used for finding an area can be extended to calculate the moment of the area. Once the latter is found, we can calculate the distance of the centroid of the area from a given axis.

This is workable because the distance of the centroid of any area (from a given axis) is equal to:

$$\frac{\text{Moment of area about an axis}}{\text{Area}}$$

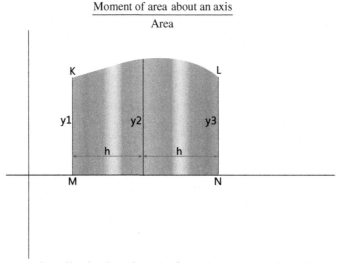

Now let us use an actual application based on the formula we saw earlier. We shall consider a ship of 160.00 m length, which is floating on a waterplane, the half breadths of which are:

Station	Half-Breadth (m)
Aft	
0	1.61
½	2.7
1	3.6
2	4.7
3	5.2
4	5.1
5	4.6
6	3.6
7	2.0
7½	1.08
8	0
For'd	

Given the information in the previous table, we are required to find the centroid of this waterplane.

Station	Half Breadths (m)	Simpson's Multiplier	Products for Areas, K	Levers	Products for Moments, L
Aft					
0 [A.P.P.]	1.61	½	0.805	4	3.220
½	2.7	2	5.400	3½	18.900
1	3.6	1½	5.400	3	16.200
2	4.7	4	18.800	2	37.600
3	5.2	2	10.400	1	10.400
4	5.1	4	20.400	0	86.320
5	4.6	2	9.200	1	9.200
6	3.6	4	14.400	2	28.800
7	2.0	1½	3.000	3	9.000
7½	1.08	2	2.160	3½	7.560
8 [F.P.P.]	0	½	0	4	0
For'd			89.965		54.560

Amidships

The common interval is equal to	$h = 160/8 = 20\,\text{m}$
The area is equal to	$K*(h/3)*2$ (2 is a multiplier that takes into account both sides of the waterplane)
	$89.965*(20/3)*2$
	$1199.533\,\text{m}^2$

The difference of the moments for'd and aft is equal to: $86.320 - 54.560 = 31.760$, with the excess being aft. The center of flotation of this particular waterplane can be found as follows:

$$\frac{\left[\dfrac{h}{3}\right]*\text{excess in moments}*h*2}{\left[\dfrac{h}{3}\right]*K*2}$$

$$= \frac{\left(\dfrac{20}{3}\right)*31.760*20*2}{\left(\dfrac{20}{3}\right)*89.965*2}$$

$$= \frac{1270.4}{179.93}$$

$$= 7.060\,\text{m, aft of amidships}$$

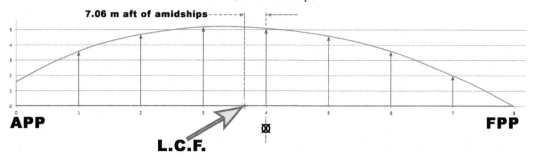

7.06 m aft of amidships

APP   L.C.F.   FPP

Since we have calculated the area of this waterplane, we can now calculate the TPC, "tonnes per centimeter immersion."

TPC is the mass that may be added (or subtracted, for that matter) for the draft to be increased (or decreased) by 1 cm.

The change of volume will be:	(Area of waterplane) * 0.01 (unit: m³)
The change of displacement will be:	(Area of waterplane) * 0.01 * 1.025 (unit: tonnes in salt water)
Accordingly, TPC:	$\dfrac{\text{Area of waterplane}}{100}$
In our case, TPC	11.99 tonnes

In order to calculate the position of the L.C.F. (longitudinal center of flotation), we have utilized Simpson's First Rule; however, this does not mean that the L.C.F. cannot be found by using other rules, such as Simpson's Second Rule or "5, 8, −1" Rule.

Further application of calculations similar to the ones we saw above and in the earlier pages the Naval Architect can calculate the volume of displacement of a vessel, which will enable him/her to calculate L.C.B. (longitudinal center of buoyancy) and progress to construct a "Displacement Table."

By varying the draft and the above calculations, we can produce curves showing:

- Extreme displacement in salt water
- V.C.B., vertical center of buoyancy
- Transverse metacentric height
- Longitudinal metacentric height
- TPC, tonnes per centimeter immersion
- MTCT 1 cm, moment to change trim 1 cm
- L.C.F., longitudinal center of flotation.
- L.C.B., longitudinal center of buoyancy.

The plotting of these curves results in what is known as a table of "Hydrostatic Curves." These tables are very important to those who design ships and to those who sail on them.

For more accurate results, all of the above data are tabulated against drafts and they are called "Hydrostatic Tables."

FREEBOARD AND LOAD LINE

To prevent damage to, or loss of merchant ships, a freeboard had to be allocated. The rules that dictate the size of the freeboard for each ship as well as other relevant details are contained within the load line convention.

The first international load line convention was adopted in 1930 and readopted in 1966, with other amendments having been made in 1971, 1975, 1979, and 1983.

Amendments were instituted in 2003. These entered into force in early 2005. The purpose of the load line marks is to provide a quick check to the observer as to whether the ship may be overloaded, or not, with a further confirmation that she still possesses sufficient freeboard.

The calculations involved here are dependent on the density of the water in which the ship floats and on her length, depth, beam, and sheer.

Other factors that must be taken into consideration are watertight superstructures (Figure 26).

The picture on the left shows the Load Line marks. The Freeboard is measured between the Summer Load Line and the line representing the level of the 'deck at side'.

Accordingly, if a vessel is loaded to her Summer Load Line she will have a freeboard as shown. These marks not only allow the observer to visually check the freeboard, but to ensure that there is enough buoyancy.

FIGURE 26

Freeboard and load line marks.

COMMON TERMS

Displacement: The water displaced by the hull of a ship is called Displacement.

The Displacement may be expressed as a volume, or as a weight, or as a mass.

The Displacement as a volume is the empty space in the water that is equal to the underwater shape of a ship.

Unit = m^3.

The Displacement as a weight is the weight of the water that is displaced by a ship.

Unit, see below.

Displacement in Fresh Water = Volume of Displacement * $1000 \, kg/m^3$.

Displacement in Sea Water = Volume of Displacement * $1025 \, kg/m^3$.

The Displacement as a mass is the water that is displaced by a ship.

Unit = tonnes.

Molded Displacement: The water that is displaced by a ship's molded dimensions when she is floating at the designed load waterline.

Unit = tonnes.

Extreme Displacement: This is the displacement calculated as:

[Molded displacement] + [shell plating displacement] + [displacement of all appendages]

Lightship Displacement: This is the extreme displacement, with the ship fully equipped, ready for sea, but without crew, stores, fuel, cargo, passengers, and water; however, the boiler(s) are filled with water to working level.

Deadweight: This is equal to [Extreme Displacement] − [Lightship Displacement].

Scantlings: The thickness (and dimensions) of the plates that make up a ship.

Parallel Middle Body: The length over which the midship section remains unchanged is called the parallel middle body.

Trim: The difference between the drafts forward and aft is called trim. When the draft forward is greater than the draft aft, the ship is said to be trimmed by the head. When the draft aft is greater than the draft forward, the ship is said to be trimmed by the stern. If the drafts are the same, forward and aft, the ship is said to float on an even keel (Figures 27 and 28).

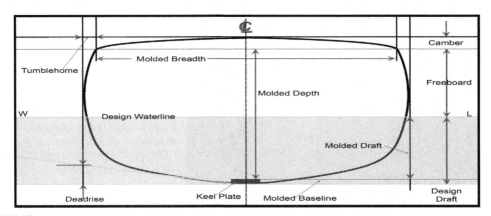

FIGURE 27

Ship's tumblehome, deadrise, and camber.

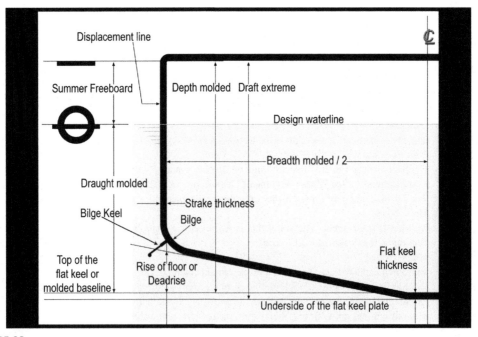

FIGURE 28

Terms relevant to a ship's midship section.

Camber: The deck of a ship, when viewed on a transverse section, may be curved. The high point of this curve lies on the centerline of the ship, and it tapers to zero towards the deck at side. Camber is also called "round of beam."

Rise of Floor: The amount the line of the outer bottom amidships rises above the baseline—when extended to the molded breadth line at side—is called "rise of floor."

Flat of Keel: The length of flat bottom plating on either side of the ship's centerline is called "flat of keel."

Bilge: The rounded plating at the corner of the vertical side shell plating and the bottom shell plating is called "bilge."

Entrance and Run: The hull underwater portion forward of the parallel middle body is called "entrance," whereas the part located aft of the parallel middle body is called "run" (Figure 29).

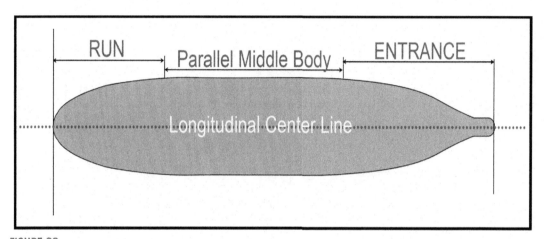

FIGURE 29

Entrance and run.

Flare: The shell plating, immediately aft of the stem, is given a concave form which, in the photo above (Figure 30), is highlighted by the curvature of the black shape. This curvature is called "flare."

Tumblehome: The fall-in of the midship section as compared to the half breadth of the ship is called tumblehome.

Heel: The inclination of a ship to port or starboard is called heel. This inclination is measured in degrees.

Forward Perpendicular: The forward perpendicular (FPP) is defined as a vertical line passing through the intersection of the designed load waterline and the forward face of the ship's stem.

After Perpendicular: The after perpendicular is defined as the vertical line passing through the intersection of the after face of the ship's rudder post and the designed load waterline (Figures 31 and 32).

Sheer: The difference between the height of the deck at side and the height of the deck at amidships.

TPC: Tonnes per centimeter immersion: A certain mass is required to be added to, or subtracted from a ship, in order to change her mean draft by 1 cm. It follows that if the waterplane area on which the ship is floating is AW square meters, then:

Change of Volume = AW $*$ 0.01 cm^3.

Change of Displacement = AW $*$ 0.01 $*$ 1.025 tonnes (in salt water).

Change of Displacement = AW $*$ 0.01 tonnes (in fresh water).

Center of Floatation: The center of gravity of the waterplane area of a ship is called the "center of flotation," "CF." When the angle of trim is small, then waterlines pass through the "CF."

FIGURE 30

Bow flare.

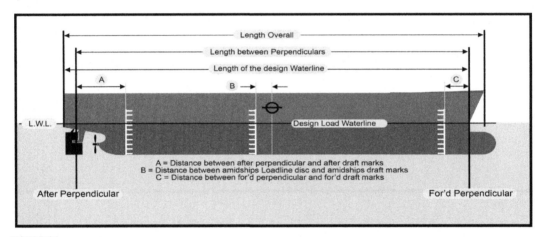

A = Distance between after perpendicular and after draft marks
B = Distance between amidships Loadline disc and amidships draft marks
C = Distance between for'd perpendicular and for'd draft marks

FIGURE 31

Ship principal dimensions.

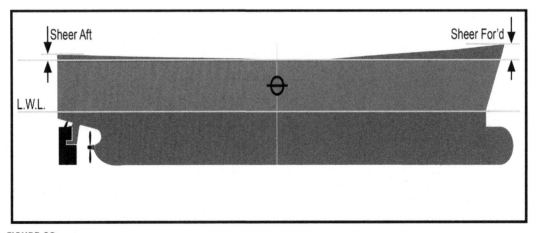

FIGURE 32

Ship's sheer.

SHIPBUILDING BASICS AND STRENGTH OF SHIPS

Ship construction, today, requires the joining of pieces of metal by way of fusion. Before we start to look at the forces acting upon metals, it is worth having a look at how the metal used in shipbuilding is produced.

Iron ore, a compound of iron and oxygen, including small amounts of other elements, is used to obtain iron. Steel is an alloy made up of iron and carbon with the inclusion of nickel and chrome. It is said that about 5% of the earth's surface is iron.

A blast furnace is utilized to produce iron from ores, and the molten product is cast into ingots or it is turned into pig iron. Today, to turn iron into steel, a basic oxygen furnace is used which allows the introduction of additional elements.

The molten steel produced through this process is cast into ingots and leaves the furnace to be stored until required. In the steel making furnace, the steel is mixed with manganese, nickel, aluminum, silicon (and other additives) which help to produce capped steel, killed steel, semi-killed steel, and rimmed steel.

The steel product composition, at this stage, begins to acquire the characteristics of mild steel. The ingots are rolled into a rectangular-shaped slab which is hot worked to produce (Figure 33):

1 Hot rolled steel plates,
2 Hot and cold rolled steel sheets, in coils,
3 Steel strips used to produce welded pipes and tubes.
4 Blooms and billets that can be hot rolled to further produce:
4a Beams,
4b Angles
4c Flats,
4d Squares,
4e Round bars,
4f Rails,
4g Reinforcing bars.

SHIPBUILDING REQUIREMENTS

There are a number of requirements involved in shipbuilding, such as savings in the length of welding lines, reduction in the hours spent welding steels, reduction in quality control costs, to mention but a few.

Shipbuilders require a reduced plate thickness, a saving in weight in the final structure, the ability to deal with heavy loads, and cost saving in weights and general fabrication processes (Figure 34).

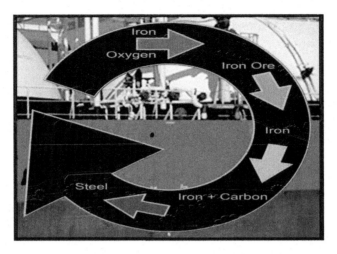

FIGURE 33

A reminder diagram of how steel is manufactured.

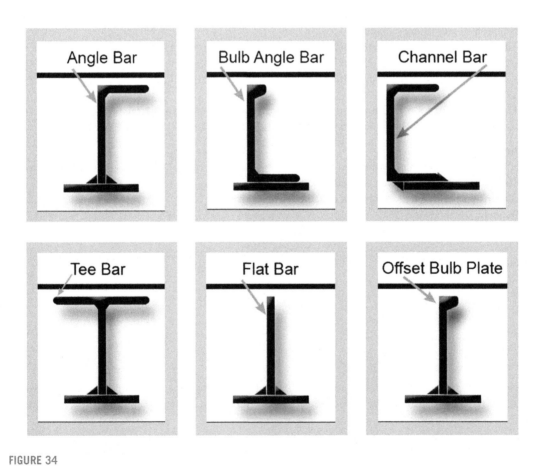

FIGURE 34

Steel sections used in shipbuilding.

The grades of steel used in shipbuilding are A, B, C, D, and E.

Grade A Ordinary mild steel
Grade B Ordinary mild steel
Grade C Steel of higher notch toughness
Grade D Steel of higher notch toughness
Grade E Steel of higher notch toughness

A number of steel components used in the building of a ship may be subjected to heat treatments such as annealing or normalizing so that the steel gains a refined grained structure.

This treatment results in an improvement of tensile strength and resistance to shock and ductility.

The forces that act on a ship structure are:

- *Stress*: A force acting per square millimeter, expressed in kilograms.
- *Strain*: An alteration in shell plating, or in a profile due to the application of stress.

A strain may be considered as under:

① *Tensile*: This force tends to elongate a steel component.
② *Compressive*: This force tends to compress a steel component.
③ *Shear*: This is a combination of two parallel forces acting in opposite directions, thus causing a "push/pull" effect.

Generally, one can look upon a force (or a load for that matter) as the form of energy that causes a change to the shape of a material—a deformation.

When the force stops acting on the material, any deformation that had taken place will disappear and the material will return to its original shape. Sometimes protracted application of a force, too large for the material to withstand, causes the material to remain deformed.

The above go to show that if a steel rod is subjected to a tensile strain it will undergo elongation; however, once this strain is removed, the rod will return to its original shape and length. So long as this return of the material to its original state keeps on happening, the material is termed to be "Elastic."

If this process is repeated with a concurrent increment of the strain applied, a point will be reached when the material is no longer capable of returning to its original state, even if the strain is removed. When this phenomenon occurs, the material is said to have reached its "yield point." The force at which this occurs is termed "yield point load."

The system of shell plating welded to beams, frames, stiffeners, brackets, stays, and stringers (to mention but a few)—provided design calculations have been carried out accurately and correctly—produces a satisfactory structural arrangement which is sufficiently strong to resist shear forces, bending moments, point loads, and UDLs that come about as effects produced by adverse weather or cargo weights, or both.

FORCES ACTING ON A SHIP

There are two main forces involved with a ship's structure when she is safely moored alongside a port berth. One of them is the sum total of all the items that go to make up her weight, and the other is the pressure exerted on all points of her hull by the water in which she floats.

Those who have been at sea, or those who inspect ships, are well aware of the fact that ships bend, and the reason of this phenomenon is because the two forces mentioned above are never equal and opposite.

The various strain forces that are applied to a ship when she is out at sea are both significant and more complex than when she is in port, causing her to sag and hog when her boundary is considered longitudinally.

Simultaneously, she is subjected to torque forces due to the irregular action of the waves on her hull. To assess strength requirements, when designing a ship, it is the usual practice to consider her longitudinal boundary as a simply supported beam, capable of flexing upwards (hogging) or downwards (sagging) (Figure 35).

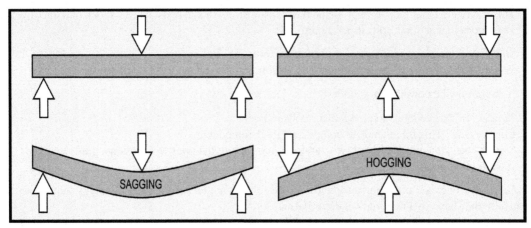

FIGURE 35

Simply supported beam, sagging, and hogging.

The beam equations, which apply in the transverse section of a ship, are:

$$\frac{M}{I} = \frac{\sigma}{y} = \frac{E}{r}$$

where:

M = bending moment
I = second moment of area
σ = bending stress
y = distance from neutral axis
E = modulus of elasticity
r = radius of curvature

The diagram in Figure 36 shows that the shear force is carried by the web, with the maximum value occurring at the neutral axis. However, the moment of resistance to bending is taken up by the flanges.

The example in Figure 37 demonstrates how the Section Modulus varies depending on whether the material is stacked vertically or horizontally. The beam on the left has a Section Modulus that is higher than that of the beam on the right. The beam on the left is stronger than the beam on the right, and it is capable of supporting higher loads.

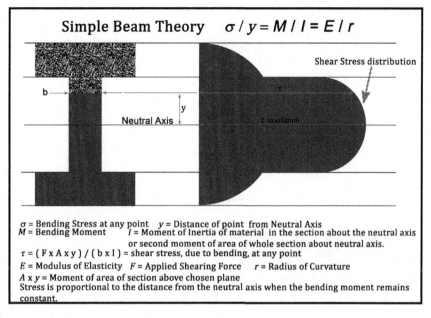

Simple Beam Theory $\sigma / y = M / I = E / r$

Shear Stress distribution

b

y

Neutral Axis

τ

τ maximum

σ = Bending Stress at any point y = Distance of point from Neutral Axis
M = Bending Moment I = Moment of Inertia of material in the section about the neutral axis
                           or second moment of area of whole section about neutral axis.
$\tau = (F \times A \times y) / (b \times I)$ = shear stress, due to bending, at any point
E = Modulus of Elasticity F = Applied Shearing Force r = Radius of Curvature
$A \times y$ = Moment of area of section above chosen plane
Stress is proportional to the distance from the neutral axis when the bending moment remains constant.

FIGURE 36

Shear stress at any point.

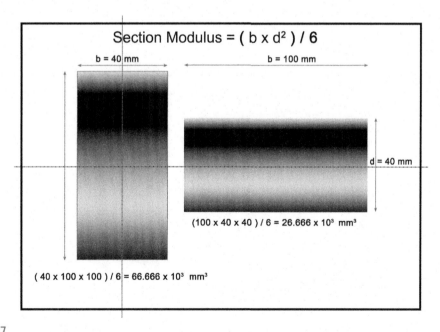

Section Modulus = (b x d²) / 6

b = 40 mm

b = 100 mm

d = 40 mm

$(100 \times 40 \times 40) / 6 = 26.666 \times 10^3$ mm³

$(40 \times 100 \times 100) / 6 = 66.666 \times 10^3$ mm³

FIGURE 37

Section Modulus calculation.

FORCES ACTING ON A SHIP AT SEA

In earlier pages, we looked at the forces of buoyancy and weight as well as those forces acting on a ship's hull by water pressure. Once the ship proceeds to sea and she is in motion, six dynamic forces act on her hull, namely (Figures 38–40):

(1) Heaving
(2) Swaying
(3) Surging
(4) Yawing
(5) Pitching
(6) Rolling

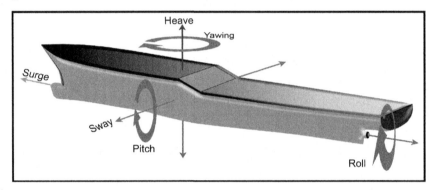

FIGURE 38

Forces acting on a ship at sea.

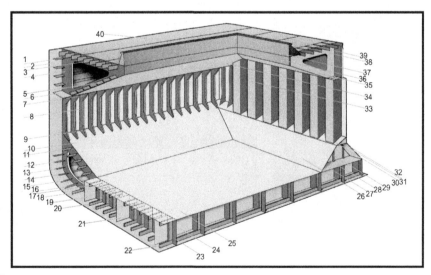

FIGURE 39

Cargo hold, single skin bulk carrier.

Source: Lloyd Register RFICL.

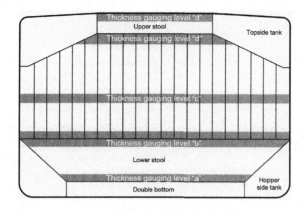

FIGURE 40

Bulkhead thickness gauging levels.

Source: IACS.

MAIN STRUCTURAL PARTS OF A SHIP

This section covers certain structural members of a hull, showing the features of each area by way of examining the particular diagrams. An inspection should aim to record affected ship parts, making sufficient differentiation between:

(1) The parts that are affected by mechanical damage.
(2) The parts that are affected by deterioration (wear and tear).
(3) The parts that are affected by external forces or by incorrect cargo loading (Table 1).

Class periodical surveys may require the taking of *thickness measurements*. The shaded strips should allow an accurate evaluation of any diminution in scantlings that may have occurred.

Level "a": Immediately above the double bottom (inner bottom) and within 30 mm from it.
Level "b": Immediately above and below the lower stool shelf plate (for those ships fitted with lower stools) and immediately above the line of the shedder plates.
Level "c": About mid-height of the bulkhead.
Level "d": At the upper part of the bulkhead adjacent to the upper deck or immediately below the upper stool shelf plate (for those ships fitted with upper stools).

This section is important because the attending surveyor must know the names of the different parts of a ship so that he/she can accurately identify them during his/her attendance and describe the extent of the damage observed.

The hatch coaming, topside tank plating, topside tank slopping plating side shell and side shell frames, hopper tank slopping plating, and inner bottom plating (tank top) will all be structural members to be scrutinized and reported on (Figures 41 and 42).

Equally, all exposed structural parts that make up the transverse bulkheads in a cargo hold, for'd and aft, will require careful inspection, including cargo hatchway end coamings and transverse beams and cross-deck structures.

Figure 43 depicts the areas where fractures occur at welded connections of the lower stool plating to the inner bottom, in the way of the duct keel. The part of the drawing on the left shows failures that are more

Table 1 Table for Figure 39			
1	Side shell longitudinal	21	Side girder
2	Side shell plating	22	Bottom longitudinal
3	Strength deck longitudinal	23	Double bottom tank floor
4	Topside tank transverse	24	Side girder
5	Topside tank sloping plating longitudinal	25	Inner bottom plating, or tanktop
6	Topside tank bottom plating	26	Lower stool transverse sloping plating
7	Side shell plating frame top bracket	27	Tanktop extension, within lower stool
8	Side shell plating frame	28	Continuous girder, within lower stool
9	Side shell plating frame bottom bracket	29	Stiffener to continuous girder, within lower stool
10	Hopper tank sloping plating	30	Shedder plate
11	Hopper tank sloping plating longitudinal	31	Shelf plate
12	Side shell longitudinal	32	Lower stool transverse sloping plating, in the adjacent hold
13	Hopper tank bilge longitudinal	33	Corrugated bulkhead
14	Hopper tank bilge longitudinal	34	Topside tank sloping plating
15	Hopper tank bilge longitudinal	35	Upper stool transverse
16	Hopper tank flat bottom longitudinal	36	Cargo hatchway transverse beam
17	Inner bottom longitudinal	37	Cargo hatchway end coaming
18	Hopper tank flat bottom longitudinal	38	Cross deck structure transverse beam
19	Side girder	39	Cross deck (structure) plating
20	Bottom longitudinal	40	Weather deck (strength deck) plating

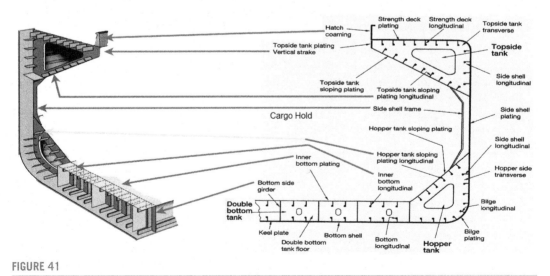

FIGURE 41

Transverse section of the bottom, side, and weather deck of a ship.

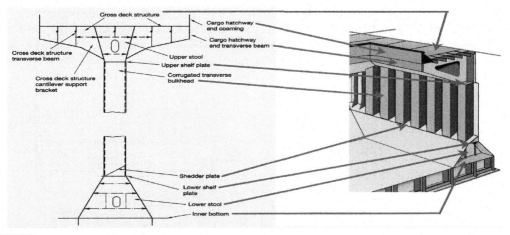

FIGURE 42

Corrugated transverse bulkhead, cross-deck structure and lower stool.

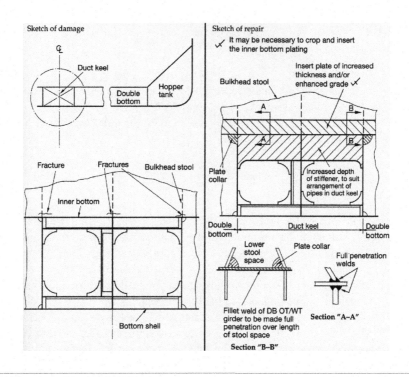

FIGURE 43

Transverse bulkheads and associated structure.

Source: IACS.

likely to occur at the boundaries of ballast holds. These fractures arise because of stress concentrations (in way of cutouts) which are exacerbated by the flexibility of the inner bottom structure i.w.o. the duct keel.

The part of the drawing on the right shows where additional reinforcements should be fitted in order to prevent recurrence of the damage presented within the part of the drawing on the left (Figures 44 and 45).

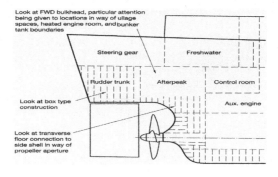

FIGURE 44

After Peak structure.

Source: IACS.

The fore peak tank is designed to trim the vessel. (The fore peak tank incorporates the bulbous bow.) The trimming of the ship is achieved by way of pumping ballast water in or out of this tank.

There are two chain lockers, one on the port and one on the starboard side, located near the collision bulkhead and at the top part of the fore peak tank.

SHIPS' CLASSIFICATION SOCIETIES AND IMO
CLASSIFICATION SOCIETIES

Edward Lloyd established his coffee house in 1688, in Tower Street, London, England. The majority of his customers were involved in the maritime world of that time. They were brokers, merchants, and underwriters, and they found Lloyd's house a good place to drink coffee and do business.

By 1733, Lloyd's business had grown to the point that the original coffee house was no longer large enough to conduct business. A new location was chosen by John Julius Angerstein (also known as the Father of Lloyd's), near the Bank of England, in the Royal Exchange.

Between 1700 and 1820, Great Britain was at war with the French and with the Americans who had decided to rebel. During these years, insurance business was extremely brisk and underwriters (*) realized that a way of assessing the quality of the ships that they were being asked to insure was necessary.

The establishment of the Register Society took place in 1760, and it was the first classification society. Later Lloyd's Register became what used to be the Register Society. The condition of ships' Hull and Equipment determined their class. Since 1760, Lloyd's Register has published the results of ships' classification annually.

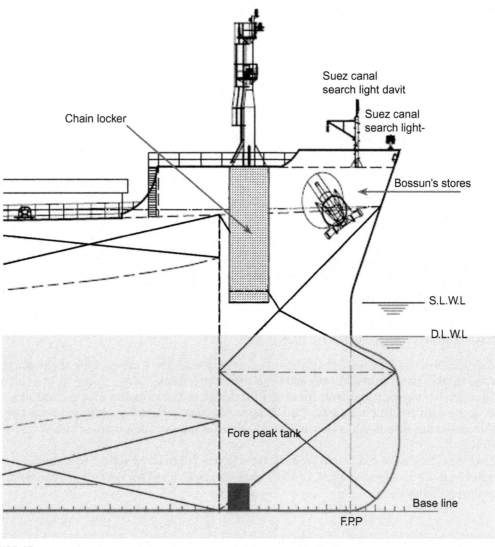

FIGURE 45

Fore peak, chain locker, and Bossun's store.

Hulls received the letters A, E, I, O, or U, depending on the state of their construction and their soundness. Ships' Equipment received the letters G (good), M (middling), or B (bad). Eventually, the numerals 1, 2, and 3 came into use instead of the letters G, M, and B. This change formed the well-known expression "A1," meaning "first or highest class."

The purpose of this system was for evaluating risk and had nothing to do with safety, fitness for purpose, or seaworthiness of the ship.

In 1834, Lloyd's Register of British and Foreign Shipping became a self-standing classification society in the United Kingdom. The first surveyors appointed at that time were retired sea captains.

Lloyd's Register created a General Committee for the running of the Society and for the publishing of the Rules regarding ship construction and maintenance.

In 1834, the Register Society published the first Rules for the survey and classification of vessels and changed its name to Lloyd's Register of Shipping. From that time onwards, surveyors and support staff became full-time employees.

Other maritime nations formed their own classification societies, to compete with Lloyd's Register. One of these was Bureau Veritas (BV), founded in Antwerp in 1828 and moving to Paris in 1832.

Others followed, like:

Registro Italiano Navale (RINA) in 1861
American Bureau of Shipping (ABS) in 1862
Det Norske Veritas (DNV) in 1864
Germanischer Lloyd (GL) in 1867
Nippon Kaiji Kyokai (ClassNK) in 1899
Russian Maritime Register of Shipping (RS) in 1913
Yugoslav Register of Shipping (now Croatia's CRS) in 1949
China Classification Society (CCS) in 1956
Korean Register (KR) in 1960
Indian Register of Shipping (IRS) in 1975

Today, the work of classification societies involves the promotion of safety of life at sea, and their rules and regulations safeguard property and protect the environment.

They develop technical standards (rules) for the design and construction of ships, and they approve designs against their standards. Classification societies hold surveys during construction of ships to ensure that their building is in accordance with the approved design requirements set out in the Rules.

In addition, they act as recognized organizations carrying out statutory surveys and certification as delegated by maritime administrations, and they carry out research and development programs related to the shipping industry.

INTERNATIONAL MARITIME ORGANIZATION (IMO)

IMO is a specialized agency of the United Nations that is responsible for measures to improve the safety and security of international shipping and to prevent marine pollution from ships. The IMO's objectives can be best summed up by its slogan—"Safe, secure, and efficient shipping on clean oceans." It was established by means of a convention adopted in Geneva in 1948 and first met in 1959. Based in the United Kingdom, the IMO has 169 member states as of 2010 and 3 associate members.

IMO is also involved in legal matters pertaining to international shipping, such as liability and compensation matters, and the facilitation of international maritime traffic. The IMO's governing body, which is the Assembly that is made up of all 169 member states, generally meets every 2 years.

SHIP'S CERTIFICATES

FAL.2/Circ.127
MEPC.1/Circ.817
MSC.1/Circ.1462
1 July 2013
IMO circular, per the above shown reference states:

List of Certificates and Documents Required to Be Carried on Board Ships, 2013

1. The Facilitation Committee, at its 38th session (8-12 April 2013); the Marine Environment Protection Committee, at its 65th session (13-17 May 2013); and the Maritime Safety Committee, at its 92nd session (12-21 June 2013) approved the list of certificates and documents required to be carried on board ships, 2013, as set out in the annex.

2. This work was carried out in accordance with the provisions of Section 2 of the annex to the FAL Convention concerning formalities required of shipowners by public authorities on the arrival, stay, and departure of ships. It is reiterated that these provisions should not read as precluding a requirement for the presentation for inspection by the appropriate authorities of certificates and other documents carried by the ship pertaining to its registry, measurement, safety, manning, classification, and other related matters.

3. Due to amendments to relevant instruments since the issuance of FAL.2/Circ.123-MEPC/Circ.769-MSC/Circ.1409, the list has been revised to take account of the relevant provisions of the aforementioned amendments.

4. This circular lists only the certificates and documents that are required under IMO instruments, and it does not include certificates or documents required by other international organizations or governmental authorities.

5. This circular should not be used in the context of port state control inspections for which convention requirements should be referred to.

6. Member Governments are invited to note the information provided in the annex and take action as appropriate.

7. This circular supersedes FAL.2/Circ.123-MEPC/Circ.769-MSC/Circ.1409. I:\CIRC\MSC\01\1462.doc

The following certificates have been displayed in the following two pages, for reference purposes:

(A) CERTIFICATE OF CLASSIFICATION (ABS)
(B) PREVENTION OF AIR POLLUTION —CERTIFICATE OF COMPLIANCE (LR) (Figures 46 and 47)

The certificates (must be originals) and documents required to be carried onboard ships are as follows:

1: All ships to which the referenced convention applies		
TITLE:	**International Tonnage Certificate (1969)**	01
REFERENCE:	**LL Convention, article 16; 1988 LL Protocol, article 16.**	
An International Tonnage Certificate (1969) shall be issued to every ship, the gross and net tonnage of which has been determined in accordance with the Convention.		

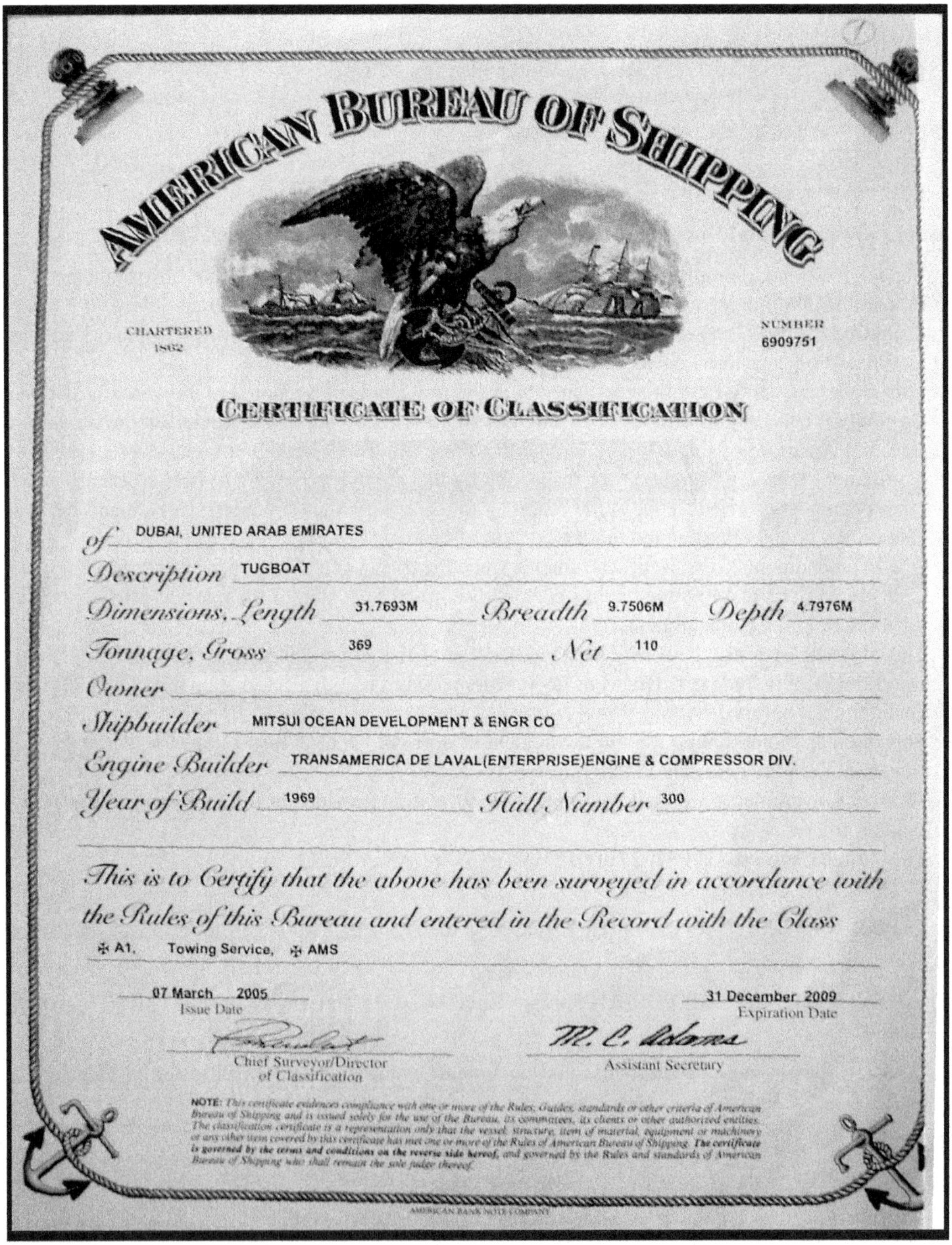

AMERICAN BUREAU OF SHIPPING

CHARTERED 1862

NUMBER 6909751

CERTIFICATE OF CLASSIFICATION

of DUBAI, UNITED ARAB EMIRATES

Description TUGBOAT

Dimensions, Length 31.7693M *Breadth* 9.7506M *Depth* 4.7976M

Tonnage, Gross 369 *Net* 110

Owner

Shipbuilder MITSUI OCEAN DEVELOPMENT & ENGR CO

Engine Builder TRANSAMERICA DE LAVAL(ENTERPRISE)ENGINE & COMPRESSOR DIV.

Year of Build 1969 *Hull Number* 300

This is to Certify that the above has been surveyed in accordance with the Rules of this Bureau and entered in the Record with the Class

✠ A1, Towing Service, ✠ AMS

07 March 2005
Issue Date

31 December 2009
Expiration Date

Chief Surveyor/Director of Classification

M. C. Adams
Assistant Secretary

NOTE: *This certificate evidences compliance with one or more of the Rules, Guides, standards or other criteria of American Bureau of Shipping and is issued solely for the use of the Bureau, its committees, its clients or other authorized entities. The classification certificate is a representation only that the vessel, structure, item of material, equipment or machinery or any other item covered by this certificate has met one or more of the Rules of American Bureau of Shipping.* ***The certificate is governed by the terms and conditions on the reverse side hereof,*** *and governed by the Rules and standards of American Bureau of Shipping who shall remain the sole judge thereof.*

AMERICAN BANK NOTE COMPANY

FIGURE 46

Certificate of Classification. ABS.

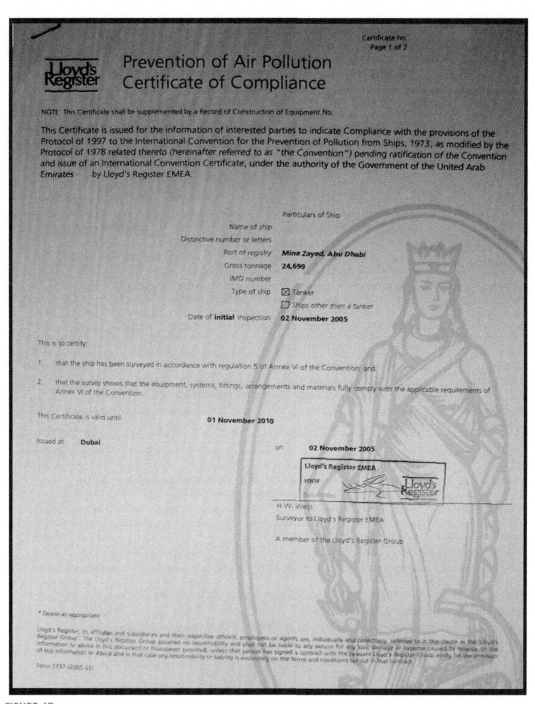

Lloyd's Register

Prevention of Air Pollution
Certificate of Compliance

NOTE: This Certificate shall be supplemented by a Record of Construction of Equipment No:

This Certificate is issued for the information of interested parties to indicate Compliance with the provisions of the Protocol of 1997 to the International Convention for the Prevention of Pollution from Ships, 1973, as modified by the Protocol of 1978 related thereto (hereinafter referred to as "the Convention") pending ratification of the Convention and issue of an International Convention Certificate, under the authority of the Government of the United Arab Emirates   by Lloyd's Register EMEA.

Particulars of Ship

Name of ship	
Distinctive number or letters	
Port of registry	**Mina Zayed, Abu Dhabi**
Gross tonnage	**24,699**
IMO number	
Type of ship	☒ Tanker
	☐ Ships other than a tanker
Date of **initial** inspection	**02 November 2005**

This is to certify:

1. that the ship has been surveyed in accordance with regulation 5 of Annex VI of the Convention; and

2. that the survey shows that the equipment, systems, fittings, arrangements and materials fully comply with the applicable requirements of Annex VI of the Convention.

This Certificate is valid until   **01 November 2010**

Issued at   **Dubai**

on   **02 November 2005**

Lloyd's Register EMEA

HWW

H.W. Weijs

Surveyor to Lloyd's Register EMEA

A member of the Lloyd's Register Group

* Delete as appropriate

Form 1737 (2005.07)

FIGURE 47

Prevention of Air Pollution Certificate of Compliance. LR.

TITLE:	**International Load Line Certificate**	02
REFERENCE:	**LL Convention, article 16; 1988 LL Protocol, article 16**	

An International Load Line Certificate shall be issued under the provisions of the International Convention on Load Lines, 1966, to every ship which has been surveyed and marked in accordance with the Convention or the Convention as modified by the 1988 LL Protocol, as appropriate.

TITLE:	**International Load Line Exemption Certificate**	03
REFERENCE:	**LL Convention, article 6; 1988 LL Protocol, article 16**	

An International Load Line Exemption Certificate shall be issued to any ship to which an exemption has been granted under and in accordance with article 6 of the Load Line Convention or the Convention as modified by the 1988 LL Protocol, as appropriate.

TITLE:	**Coating Technical File**	04
REFERENCE:	**SOLAS 1974, regulation II-1/3-2; Performance standard for protective coatings for dedicated seawater ballast tanks in all types of ships and double-side skin spaces of bulk carriers (resolution MSC.215 (82))**	

A Coating Technical File, containing specifications of the coating system applied to dedicated seawater ballast tanks in all types of ships and double-side skin spaces of bulk carriers of 150 m in length and upwards, a record of the shipyard's and shipowner's coating work, detailed criteria for coating sections, job specifications, inspection, maintenance and repair, shall be kept on board and maintained throughout the life of the ship.

TITLE:	**Construction drawings**	05
REFERENCE:	**SOLAS 1974, regulation II-1/3-7; MSC/Circ.1135 on As-built construction drawings to be maintained on board the ship and ashore**	

A set of as-built construction drawings and other plans showing any subsequent structural alterations shall be kept on board a ship constructed on or after 1 January 2007.

TITLE:	**Ship Construction File**	06
REFERENCE:	**SOLAS 1974, regulation II-1/3-10; MSC.1/Circ.1343 on Guidelines for the information to be included in a Ship Construction File**	

A Ship Construction File with specific information should be kept on board oil tankers of 150 m in length and above and bulk carriers of 150 m in length and above, constructed with a single deck, topside tanks and hopper side tanks in cargo spaces, excluding ore carriers and combination carriers:

.1 for which the building contract is placed on or after 1 July 2016;

.2 in the absence of a building contract, the keels of which are laid or which are at a similar stage of construction on or after 1 July 2017; or

.3 the delivery of which is on or after 1 July 2020 shall carry a Ship Construction File containing information in accordance with regulations and guidelines,

and updated as appropriate throughout the ship's life in order to facilitate safe operation, maintenance, survey, repair and emergency measures.

| TITLE: | Stability information | 07 |
| REFERENCE: | SOLAS 1974, regulations II-1/5 and II-1/5-1; LL Convention; 1988 LL Protocol, regulation 10 | |

Every passenger ship regardless of size and every cargo ship of 24 m and over shall be inclined on completion and the elements of their stability determined. The master shall be supplied with stability information containing such information as is necessary to enable him, by rapid and simple procedures, to obtain accurate guidance as to the stability of the ship under varying conditions of service to maintain the required intact stability and stability after damage. For bulk carriers, the information required in a bulk carrier booklet may be contained in the stability information.

| TITLE: | Damage control plans and booklets | 08 |
| REFERENCE: | SOLAS 1974, regulation II-1/19; MSC.1/Circ.1245 | |

On passenger and cargo ships, there shall be permanently exhibited plans showing clearly for each deck and hold the boundaries of the watertight compartments, the openings therein with the means of closure and position of any controls thereof, and the arrangements for the correction of any list due to flooding. Booklets containing the aforementioned information shall be made available to the officers of the ship.

| TITLE: | Minimum safe manning document | 09 |
| REFERENCE: | SOLAS 1974, regulation V/14.2 | |

Every ship to which chapter I of the Convention applies shall be provided with an appropriate, safe manning document or equivalent issued by the Administration as evidence of the minimum safe manning.

| TITLE: | Fire safety training manual | 10 |
| REFERENCE: | SOLAS 1974, regulation II-2/15.2.3 | |

A training manual shall be written in the working language of the ship and shall be provided in each crew mess room and recreation room or in each crew cabin. The manual shall contain the instructions and information required in regulation II-2/15.2.3.4. Part of such information may be provided in the form of audio-visual aids in lieu of the manual.

| TITLE: | Fire control plan/booklet | 11 |
| REFERENCE: | SOLAS 1974, regulations II-2/15.2.4 and II-2/15.3.2 | |

General arrangement plans shall be permanently exhibited for the guidance of the ship's officers, showing clearly for each deck the control stations, the various fire sections together with particulars of the fire detection and fire alarm systems and the fire-extinguishing appliances, etc. Alternatively, at the discretion of the Administration, the aforementioned details may be set out in a booklet, a copy of which shall be supplied to each officer, and one copy shall at all times be available on board in an accessible position. Plans and booklets shall be kept up to date; any alterations shall be recorded as soon as practicable. A duplicate set of fire control plans or a booklet containing such plans shall be permanently stored in a prominently marked weathertight enclosure outside the deckhouse for the assistance of shoreside fire-fighting personnel.

| TITLE: | Onboard training and drills record | 12 |
| REFERENCE: | SOLAS 1974, regulation II-2/15.2.2.5 | |

Fire drills shall be conducted and recorded in accordance with the provisions of regulations III/19.3 and III/19.5.

TITLE:	Fire safety operational booklet	13
REFERENCE:	SOLAS 1974, regulation II-2/16.2	

The fire safety operational booklet shall contain the necessary information and instructions for the safe operation of the ship and cargo handling operations in relation to fire safety. The booklet shall be written in the working language of the ship and be provided in each crew mess room and recreation room or in each crew cabin. The booklet may be combined with the fire safety training manuals required in regulation II-2/15.2.3.

TITLE:	Maintenance plans	14
REFERENCE:	SOLAS 1974, regulations II-2/14.2.2 and II-2/14.4	

The maintenance plan shall include the necessary information about fire protection systems and fire-fighting systems and appliances as required under regulation II-2/14.2.2. For tankers, additional requirements are referred to in regulation II-2/14.4.

TITLE:	Training manual	15
REFERENCE:	SOLAS 1974, regulation III/35	

The training manual, which may comprise several volumes, shall contain instructions and information, in easily understood terms illustrated wherever possible, on the life-saving appliances provided in the ship and on the best methods of survival. Any part of such information may be provided in the form of audio-visual aids in lieu of the manual.

TITLE:	Nautical charts and nautical publications	16
REFERENCE:	SOLAS 1974, regulations V/19.2.1.4 and V/27	

Nautical charts and nautical publications for the intended voyage shall be adequate and up to date. An electronic chart display and information system (ECDIS) is also accepted as meeting the chart carriage requirements of this subparagraph.

TITLE:	International Code of Signals and a copy of Volume III of IAMSAR Manual	17
REFERENCE:	SOLAS 1974, regulation V/21	

All ships required to carry a radio installation shall carry the International Code of Signal; all ships shall carry an up-to-date copy of Volume III of the International Aeronautical and Maritime Search and Rescue (IAMSAR) Manual.

TITLE:	Records of navigational activities	18
REFERENCE:	SOLAS 1974, regulations V/26 and V/28.1	

All ships engaged on international voyages shall keep on board a record of navigational activities and incidents, including drills and pre-departure tests. When such information is not maintained in the ship's logbook, it shall be maintained in another form approved by the Administration.

TITLE:	Manoeuvring booklet	19
REFERENCE:	SOLAS 1974, regulation II-1/28	

The stopping times, ship headings and distances recorded on trials, together with the results of trials to determine the ability of ships having multiple propellers to navigate and maneuver with one or more propellers inoperative, shall be available on board for the use of the master or designated personnel.

TITLE:	**Material Safety Data Sheets (MSDS)**	20
REFERENCE:	**SOLAS 1974, regulation VI/5-1; resolution MSC.286 (86)**	

Ships carrying oil or oil fuel, as defined in regulation 1 of annex 1 of the International Convention for the Prevention of Pollution from Ships, 1973, as modified by the Protocol of 1978 relating thereto, shall be provided with material safety data sheets, based on the recommendations developed by the Organization, prior to the loading of such oil as cargo in bulk or bunkering of oil fuel.

TITLE:	**AIS test report**	21
REFERENCE:	**SOLAS 1974, regulation V/18.9; MSC.1/Circ.1252**	

The Automatic Identification System (AIS) shall be subjected to an annual test by an approved surveyor or an approved testing or servicing facility. A copy of the test report shall be retained on board and should be in accordance with a model form set out in the annex to MSC.1/Circ.1252

TITLE:	**Certificates for masters, officers or ratings**	22
REFERENCE:	**STCW 1978, article VI, regulation I/2; STCW Code, section A-I/2**	

Certificates for masters, officers or ratings shall be issued to those candidates who, to the satisfaction of the Administration, meet the requirements for service, age, medical fitness, training, qualifications and examinations in accordance with the provisions of the STCW Code annexed to the International Convention on Standards of Training, Certification and Watchkeeping for Seafarers, 1978. Formats of certificates are given in section A-I/2 of the STCW Code. Certificates must be kept available in their original form on board the ships on which the holder is serving.

TITLE:	**Records of hours of rest**	23
REFERENCE:	**STCW Code, section A-VIII/1; Maritime Labour Convention, 2006;**	
	Seafarers' Hours of Work and the Manning of Ships Convention, 1996 (No.180); IMO/ILO Guidelines for the development of tables of seafarers' shipboard working arrangements and formats of records of seafarers' hours of work or hours of rest	
	Note: The Maritime Labour Convention, 2006 shall come into force on 20/08/2013.	

Records of daily hours of rest of seafarers shall be maintained on board.

TITLE:	**International Oil Pollution Prevention Certificate**	24
REFERENCE:	**MARPOL Annex I, regulation 7**	

An international Oil Pollution Prevention Certificate shall be issued, after survey in accordance with regulation 6 of Annex I of MARPOL, to any oil tanker of 150 gross tonnage and above and any other ship of 400 gross tonnage and above which is engaged in voyages to ports or offshore terminals under the jurisdiction of other Parties to MARPOL. The certificate is supplemented with a Record of Construction and Equipment for Ships other than Oil Tankers (Form A) or a Record of Construction and Equipment for Oil Tankers (Form B), as appropriate.

TITLE:	**Oil Record Book**	25
REFERENCE:	**MARPOL Annex I, regulations 17 and 36**	

Every oil tanker of 150 gross tonnage and above and every ship of 400 gross tonnage and above other than an oil tanker shall be provided with an Oil Record Book, Part I (Machinery space operations). Every oil tanker of 150 gross tonnage and above shall also be provided with an Oil Record Book, Part II (Cargo/ballast operations).

TITLE:	**Shipboard Oil Pollution Emergency Plan**	26
REFERENCE:	**MARPOL Annex I, regulation 37; resolution MEPC.54 (32), as amended by resolution MEPC.86 (44)**	

Every oil tanker of 150 gross tonnage and above and every ship other than an oil tanker of 400 gross tonnage and above shall carry on board a Shipboard Oil Pollution Emergency Plan approved by the Administration.

TITLE:	**International Sewage Pollution Prevention Certificate**	27
REFERENCE:	**MARPOL Annex IV, regulation 5; MEPC/Circ.408**	

An International Sewage Pollution Prevention Certificate shall be issued, after an initial or renewal survey in accordance with the provisions of regulation 4 of Annex IV of MARPOL, to any ship which is required to comply with the provisions of that Annex and is engaged in voyages to ports or offshore terminals under the jurisdiction of other Parties to the Convention.

TITLE:	**Garbage Management Plan**	28
REFERENCE:	**MARPOL Annex V, regulation 10; resolution MEPC.71 (38); MEPC/Circ.317**	

Every ship of 100 gross tonnage and above and every ship which is certified to carry 15 persons or more shall carry a garbage management plan which the crew shall follow.

TITLE:	**Garbage Record Book**	29
REFERENCE:	**MARPOL Annex V, regulation 10**	

Every ship of 400 gross tonnage and above and every ship which is certified to carry 15 persons or more engaged in voyages to ports or offshore terminals under the jurisdiction of other Parties to the Convention and every fixed and floating platform engaged in exploration and exploitation of the seabed shall be provided with a Garbage Record Book.

TITLE:	**Voyage data recorder system – certificate of compliance**	30
REFERENCE:	**SOLAS 1974, regulation V/18.8**	

The voyage data recorder system, including all sensors, shall be subjected to an annual performance test. The test shall be conducted by an approved testing or servicing facility to verify the accuracy, duration and recoverability of the recorded data. In addition, tests and inspections shall be conducted to determine the serviceability of all protective enclosures and devices fitted to aid location. A copy of the certificate of compliance issued by the testing facility, stating the date of compliance and the applicable performance standards, shall be retained on board the ship.

TITLE:	**Cargo Securing Manual**	31
REFERENCE:	**SOLAS 1974, regulations VI/5.6 and VII/5; MSC.1/Circ.1353**	

All cargoes other than solid and liquid bulk cargoes, cargo units and cargo transport units, shall be loaded, stowed and secured throughout the voyage in accordance with the Cargo Securing Manual approved by the Administration. In ships with ro-ro spaces, as defined in regulation II-2/3.41, all securing of such cargoes, cargo units and cargo transport units, in accordance with the Cargo Securing Manual, shall be completed before the ship leaves the berth. The Cargo Securing Manual is required on all types of ships engaged in the carriage of all cargoes other than solid and liquid bulk cargoes, which shall be drawn up to a standard at least equivalent to the guidelines developed by the Organization.

TITLE:	Document of Compliance	32
REFERENCE:	SOLAS 1974, regulation IX/4; ISM Code, paragraph 13	

A document of compliance shall be issued to every company which complies with the requirements of the ISM Code. A copy of the document shall be kept on board.

TITLE:	Safety Management Certificate	33
REFERENCE:	SOLAS 1974, regulation IX/4; ISM Code, paragraph 13	

A Safety Management Certificate shall be issued to every ship by the Administration or an organization recognized by the Administration. The Administration or an organization recognized by it shall, before issuing the Safety Management Certificate, verify that the company and its shipboard management operate in accordance with the approved safety management system.

TITLE:	International Ship Security Certificate (ISSC) or Interim International Ship Security Certificate	34
REFERENCE:	SOLAS 1974, regulation XI-2/9.1.1; ISPS Code, part A, section 19 and appendices.	

An International Ship Security Certificate (ISSC) shall be issued to every ship by the Administration or an organization recognized by it to verify that the ship complies with the maritime security provisions of SOLAS chapter XI-2 and part A of the ISPS Code. An interim ISSC may be issued under the ISPS Code, part A, section 19.4.

TITLE:	Ship Security Plan and associated records	35
REFERENCE:	SOLAS 1974, regulation XI-2/9; ISPS Code. part A, sections 9 and 10	

Each ship shall carry on board a ship security plan approved by the Administration. The plan shall make provisions for the three security levels as defined in part A of the ISPS Code. Records of the following activities addressed in the ship security plan shall be kept on board for at least the minimum period specified by the Administration:

.1 training, drills and exercises;

.2 security threats and security incidents;

.3 breaches of security;

.4 changes in security level;

.5 communications relating to the direct security of the ship such as specific threats to the ship or to port facilities the ship is, or has been, in;

.6 internal audits and reviews of security activities;

.7 periodic review of the ship security assessment;

.8 periodic review of the ship security plan;

.9 implementation of any amendments to the plan; and

.10 maintenance, calibration and testing of any security equipment provided on board, including testing of the ship security alert system.

TITLE:	Continuous Synopsis Record (CSR)	36
REFERENCE:	SOLAS 1974, regulation XI-1/5	

Every ship to which chapter I of the Convention applies shall be issued with a Continuous Synopsis Record. The Continuous Synopsis Record provides an onboard record of the history of the ship with respect to the information recorded therein.

TITLE:	International Anti-fouling System Certificate	37
REFERENCE:	AFS Convention, regulation 2(1) of annex 4	

Ships of 400 GT and above engaged in international voyages, excluding fixed or floating platforms, FSUs, and FPSOs, shall be issued after inspection and survey an International Anti-fouling System Certificate together with a Record of Anti-fouling Systems.

TITLE:	Declaration on Anti-fouling System	38
REFERENCE:	AFS Convention, regulation 5 (1) of annex 4	

Ships of 24 m or more in length, but less than 400 GT engaged in international voyages, excluding fixed or floating platforms, FSUs, and FPSOs, shall carry a declaration signed by the owner or owner's authorized agents. Such a declaration shall be accompanied by appropriate documentation (such as a paint receipt or a contractor invoice) or contain an appropriate endorsement.

TITLE:	International Air Pollution Prevention Certificate	39
REFERENCE:	MARPOL Annex VI, regulation 6	

Ships constructed before the date of entry into force of the Protocol of 1997 shall be issued with an International Air Pollution Prevention Certificate. Any ship of 400 gross tonnage and above engaged in voyages to ports or offshore terminals under the jurisdiction of other Parties and platforms and drilling rigs engaged in voyages to waters under the sovereignty or jurisdiction of other Parties to the Protocol of 1997 shall be issued with an International Air Pollution Prevention Certificate.

TITLE:	International Energy Efficiency Certificate	40
REFERENCE:	MARPOL Annex VI, regulation 6	

An International Energy Efficiency Certificate for the ship shall be issued after a survey in accordance with the provisions of regulation 5.4 to any ships of 400 gross tonnage and above before that ship may engage in voyages to ports or offshore terminals under the jurisdiction of other Parties.

TITLE:	Ozone-depleting Substances Record Book	41
REFERENCE:	MARPOL Annex VI, regulation 12.6	

Each ship subject to MARPOL Annex VI, regulation 6.1 that has rechargeable systems that contain ozone-depleting substances shall maintain an ozone-depleting substances record book.

TITLE:	Fuel Oil Changeover Procedure and Logbook (record of fuel changeover)	42
REFERENCE:	MARPOL Annex VI, regulation 14.6	

Those ships using separate fuel oils to comply with MARPOL Annex VI, regulation 14.3 and entering or leaving an emission control area shall carry a written procedure showing how the fuel oil changeover is to be done. The volume of low-sulfur fuel oils in each tank as well as the date, time and position of the ship when any fuel oil changeover operation is completed prior to the entry into an emission control area or commenced after exit from such an area shall be recorded in such logbook as prescribed by the Administration.

TITLE:	Manufacturer's Operating Manual for Incinerators	43
REFERENCE:	MARPOL Annex VI, regulation 16.7	

Incinerators installed in accordance with the requirements of MARPOL Annex VI, regulation 16.6.1 shall be provided with a Manufacturer's Operating Manual, which is to be retained with the unit.

TITLE:	**Bunker Delivery Note and Representative Sample**	44
REFERENCE:	**MARPOL Annex VI, regulations 18.6 and 18.8.1**	

Bunker Delivery Note and representative sample of the fuel oil delivered shall be kept on board in accordance with requirements of MARPOL Annex VI, regulations 18.6 and 18.8.1.

TITLE:	**Ship Energy Efficiency Management Plan (SEEMP)**	45
REFERENCE:	**MARPOL Annex VI, regulation 22; MEPC.1/Circ.795**	

All ships of 400 gross tonnage and above, excluding platforms (including FPSOs and FSUs) and drilling rigs, regardless of their propulsion, shall keep on board a ship specific Ship Energy Efficiency Management Plan (SEEMP). This may form part of the ship's Safety management System (SMS).

TITLE:	**EEDI Technical File**	46
REFERENCE:	**MARPOL Annex VI, regulation 20**	

Applicable to ships falling into one or more of the categories in MARPOL Annex VI, regulations 2.25 to 2.35.

TITLE:	**Technical File**	47
REFERENCE:	**NOx Technical Code, paragraph 2.3.4**	

Every marine diesel engine installed on board a ship shall be provided with a Technical File. The Technical File shall be prepared by the applicant for engine certification and approved by the Administration, and is required to accompany an engine throughout its life on board ships. The Technical File shall contain the information as specified in paragraph 2.4.1 of the NOx Technical Code.

TITLE:	**Record Book of Engine Parameters**	48
REFERENCE:	**NOx Technical Code, paragraph 2.3.7**	

Where the Engine Parameter Check method in accordance with paragraph 6.2 of the NOX Technical Code is used to verify compliance, if any adjustments or modifications are made to an engine after its pre-certification, a full record of such adjustments or modifications shall be recorded in the engine's Record Book of Engine Parameters.

TITLE:	**Exemption Certificate[1]**	49
REFERENCE:	**SOLAS 1974, regulation I/12; 1988 SOLAS Protocol, regulation I/12**	

When an exemption is granted to a ship under and in accordance with the provisions of SOLAS 1974, a certificate called an Exemption Certificate shall be issued in addition to the certificates listed above.

TITLE:	**LRIT conformance test report**	50
REFERENCE:	**SOLAS 1974, regulation V/19-1; MSC.1/Circ.1307**	

A Conformance test report should be issued, on satisfactory completion of a conformance test, by the Administration or the ASP who conducted the test acting on behalf of the Administration, and should be in accordance with the model set out in appendix 2 of MSC.1/Circ.1307.

TITLE:	Noise Survey Report	51
REFERENCE:	SOLAS 1974, regulation II-1/3-12; Code on noise levels on board ships, section 4.3	
	Note: The above mandatory requirements are expected to enter into force on 1/7/2014	

Applicable to new ships of 1,600 gross tonnage and above, excluding dynamically supported crafts, high-speed crafts, fishing vessels, pipe-laying barges, crane barges, mobile offshore drilling units, pleasure yachts not engaged in trade, ships of war and troopships, ships not propelled by mechanical means, pile driving vessels and dredgers.

A noise survey report shall always be carried on board and be accessible for the crew.

For existing ships, refer to section "Other certificates and documents which are not mandatory – Noise Survey Report" (resolution A.468 (XII).

TITLE:	Ship-specific Plans and Procedures for Recovery of Persons from the Water	52
REFERENCE:	SOLAS 1974 regulation, III/17-1; Resolution MSC.346 (91); MSC.1/Circ.1447	
	Note: The above mandatory requirements are expected to enter into force on 1/7/2014	

All ships shall have ship-specific plans and procedures for recovery of persons from the water. Ships constructed before 1 July 2014 shall comply with this requirement by the first periodical or renewal safety equipment survey of the ship to be carried out after 1 July 2014, whichever comes first.

Ro-ro passenger ships which comply with regulation III/26.4 shall be deemed to comply with this regulation.

The Plans and Procedures should be considered as a part of the emergency preparedness plan required by paragraph 8 of the ISM Code.

2: In addition to the certificates listed in section 1 above, passenger ships shall carry:		
TITLE:	Passenger Ship Safety Certificate	53
REFERENCE:	SOLAS 1974, regulation I/12; 1988 SOLAS Protocol, regulation I/12	

A certificate called a Passenger Ship Safety Certificate shall be issued after inspection and survey to a passenger ship which complies with the requirements of chapters II-1, II-2, III, IV and V and any other relevant requirements of SOLAS 1974. A Record of Equipment for the Passenger Ship Safety Certificate (Form P) shall be permanently attached.

TITLE:	Special Trade Passenger Ship Safety Certificate, Special Trade Passenger Ship Space Certificate	54
REFERENCE:	STP 71, rule 5	
	SSTP 73, rule 5	

A Special Trade Passenger Ship Safety Certificate issued under the provisions of the Special Trade Passenger Ships Agreement, 1971.

A certificate called a Special Trade Passenger Ship Space Certificate shall be issued under the provisions of the Protocol on Space Requirements for Special Trade Passenger Ships, 1973.

TITLE:	Search and rescue cooperation plan	55
REFERENCE:	SOLAS 1974, regulation V/7.3	

Passenger ships to which chapter I of the Convention applies shall have on board a plan for cooperation with appropriate search and rescue services in event of an emergency.

TITLE:	List of operational limitations	56
REFERENCE:	SOLAS 1974, regulation V/30	

Passenger ships to which chapter I of the Convention applies shall keep on board a list of all limitations on the operation of the ship, including exemptions from any of the SOLAS regulations, restrictions in operating areas, weather restrictions, sea state restrictions, restrictions in permissible loads, trim, speed and any other limitations, whether imposed by the Administration or established during the design or the building stages.

TITLE:	Decision support system for masters	57
REFERENCE:	SOLAS 1974, regulation III/29	

In all passenger ships, a decision support system for emergency management shall be provided on the navigation bridge.

3: In addition to the certificates listed in section 1 above, cargo ships shall carry:

TITLE:	Cargo Ship Safety Construction Certificate	58
REFERENCE:	SOLAS 1974, regulation I/12; 1988 SOLAS Protocol, regulation I/12	

A certificate called a Cargo Ship Safety Construction Certificate shall be issued after survey to a cargo ship of 500 gross tonnage and over which satisfies the requirements for cargo ships on survey, set out in regulation I/10 of SOLAS 1974, and complies with the applicable requirements of chapters II-1 and II-2, other than those relating to fire-extinguishing appliances and fire-control plans.

TITLE:	Cargo Ship Safety Equipment Certificate	59
REFERENCE:	SOLAS 1974, regulation I/12; 1988 SOLAS Protocol, regulation I/12	

A certificate called a Cargo Ship Safety Equipment Certificate shall be issued after survey to a cargo ship of 500 gross tonnage and over which complies with the relevant requirements of chapters II-1 and II-2, III and V and any other relevant requirements of SOLAS 1974. A Record of Equipment for the Cargo Ship Safety Equipment Certificate (Form E) shall be permanently attached.

TITLE:	Cargo Ship Safety Radio Certificate	60
REFERENCE:	SOLAS 1974, regulation I/12, as amended by the GMDSS amendments;	
	1988 SOLAS Protocol, regulation I/12	

A certificate called a Cargo Ship Safety Radio Certificate shall be issued after survey to a cargo ship of 300 gross tonnage and over, fitted with a radio installation, including those used in life-saving appliances, which complies with the requirements of chapter IV and any other relevant requirements of SOLAS 1974. A Record of Equipment for the Cargo Ship Safety Radio Certificate (Form R) shall be permanently attached.

TITLE:	Cargo Ship Safety Certificate	61
REFERENCE:	1988 SOLAS Protocol, regulation I/12	

A certificate called a Cargo Ship Safety Certificate may be issued after survey to a cargo ship which complies with the relevant requirements of chapters II-1, II-2, III, IV and V and other relevant requirements of SOLAS 1974 as modified by the 1988 SOLAS Protocol, as an alternative to the Cargo Ship Safety Construction Certificate, Cargo Ship Safety Equipment Certificate and Cargo Ship Safety Radio Certificate. A Record of Equipment for the Cargo Ship Safety Certificate (Form C) shall be permanently attached.

TITLE:	**Document of authorization for the carriage of grain and grain loading manual**	62
REFERENCE:	**SOLAS 1974, regulation VI/9; International Code for the Safe Carriage of Grain in Bulk, section 3**	

A document of authorization shall be issued for every ship loaded in accordance with the regulations of the International Code for the Safe Carriage of Grain in Bulk. The document shall accompany or be incorporated into the grain loading manual provided to enable the master to meet the stability requirements of the Code.

TITLE:	**Certificate of insurance or other financial security in respect of civil liability for oil pollution damage**	63
REFERENCE:	**CLC 1969, article VII**	

A certificate attesting that insurance or other financial security is in force shall be issued to each ship carrying more than 2,000 tonnes of oil in bulk as cargo. It shall be issued or certified by the appropriate authority of the State of the ship's registry after determining that the requirements of article VII, paragraph 1, of the CLC Convention have been complied with.

TITLE:	**Certificate of insurance or other financial security in respect of civil liability for bunker oil pollution damage**	64
REFERENCE:	**Bunker Convention 2001, article 7**	

Certificate attesting that insurance or other financial security is in force in accordance with the provisions of this Convention shall be issued to each ship of greater than 1,000 GT after the appropriate authority of a State Party has determined that the requirements of article 7, paragraph 1 have been complied with. With respect to a ship registered in a State Party such certificate shall be issued or certified by the appropriate authority of the State of the ship's registry; with respect to a ship not registered in a State Party it may be issued or certified by the appropriate authority of any State Party. A State Party may authorize either an institution or an organization recognized by it to issue the certificate referred to in paragraph 2.

TITLE:	**Certificate of insurance or other financial security in respect of civil liability for oil pollution damage**	65
REFERENCE:	**CLC 1992, article VII**	

A certificate attesting that insurance or other financial security is in force in accordance with the provisions of the 1992 CLC Convention shall be issued to each ship carrying more than 2,000 tonnes of oil in bulk as cargo after the appropriate authority of a Contracting State has determined that the requirements of article VII, paragraph 1, of the Convention have been complied with. With respect to a ship registered in a Contracting State, such certificate shall be issued by the appropriate authority of the State of the ship's registry; with respect to a ship not registered in a Contracting State, it may be issued or certified by the appropriate authority of any Contracting State.

TITLE:	**Enhanced survey report file**	66
REFERENCE:	**SOLAS 1974, regulation XI-1/2; resolution A.744 (18)**	
	Note: The 2011 ESP Code is expected to come into force on 1/1/2014 and to supersede resolution A.744 (18)	

Bulk carriers and oil tankers shall have a survey report file and supporting documents complying with paragraphs 6.2 and 6.3 of Annex A and Annex B of resolution A.744 (18) – Guidelines on the enhanced programme of inspections during surveys of bulk carriers and oil tankers.

Note: refer to requirements of survey report file and supporting documents for bulk carriers and oil tankers as referred to in paragraphs 6.2 and 6.3 of annex A/annex B, part A/part B, 2011 ESP Code.

| TITLE: | Record of oil discharge monitoring and control system for the last ballast voyage | 67 |
| REFERENCE: | MARPOL Annex I, regulation 31 | |

Subject to the provisions of paragraphs 4 and 5 of regulation 3 of MARPOL Annex I, every oil tanker of 150 gross tonnage and above shall be equipped with an oil discharge monitoring and control system approved by the Administration. The system shall be fitted with a recording device to provide a continuous record of the discharge in litres per nautical mile and total quantity discharged, or the oil content and rate of discharge. The record shall be identifiable as to time and date and shall be kept for at least three years.

| TITLE: | Oil Discharge Monitoring and Control (ODMC) Operational Manual | 68 |
| REFERENCE: | MARPOL Annex I, regulation 31; resolution A.496(XII); resolution A.586(14); resolution MEPC.108(49) | |

Every oil tanker fitted with an Oil Discharge Monitoring and Control system shall be provided with instructions as to the operation of the system in accordance with an operational manual approved by the Administration.

| TITLE: | Cargo Information | 69 |
| REFERENCE: | SOLAS 1974, regulations VI/2 and XII/10; MSC/Circ.663 | |

The shipper shall provide the master or his representative with appropriate information, confirmed in writing, on the cargo, in advance of loading. In bulk carriers, the density of the cargo shall be provided in the above information.

| TITLE: | Ship Structure Access Manual | 70 |
| REFERENCE: | SOLAS 1974, regulation II-1/3-6 | |

This regulation applies to oil tankers of 500 gross tonnage and over and bulk carriers, as defined in regulation IX/1, of 20,000 gross tonnage and over, constructed on or after 1 January 2006. A ship's means of access to carry out overall and close-up inspections and thickness measurements shall be described in a Ship structure access manual approved by the Administration, an updated copy of which shall be kept on board.

| TITLE: | Bulk Carrier Booklet | 71 |
| REFERENCE: | SOLAS 1974, regulations VI/7 and XII/8; Code of Practice for the Safe Loading and Unloading of Bulk Carriers (BLU Code) | |

To enable the master to prevent excessive stress in the ship's structure, the ship loading and unloading solid bulk cargoes shall be provided with a booklet referred to in SOLAS regulation VI/7.2. The booklet shall be endorsed by the Administration or on its behalf to indicate that SOLAS regulations XII/4, 5, 6 and 7, as appropriate, are complied with. As an alternative to a separate booklet, the required information may be contained in the intact stability booklet.

| TITLE: | Crude Oil Washing Operation and Equipment Manual (COW Manual) | 72 |
| REFERENCE: | MARPOL Annex I, regulation 35; resolution MEPC.81(43) | |

Every oil tanker operating with crude oil washing systems shall be provided with an Operations and Equipment Manual detailing the system and equipment and specifying operational procedures. Such a Manual shall be to the satisfaction of the Administration and shall contain all the information set out in the specifications referred to in regulation 35 of Annex I of MARPOL.

TITLE:	Condition Assessment Scheme (CAS) Statement of Compliance, CAS Final Report and Review Record	73
REFERENCE:	MARPOL Annex I, regulations 20 and 21, Resolution MEPC.94 (46); Resolution MEPC.99 (48); Resolution 112 (50); Resolution MEPC.131 (53); Resolution MEPC.155 (55).	

A Statement of Compliance shall be issued by the Administration to every oil tanker which has been surveyed in accordance with the requirements of the Condition Assessment Scheme (CAS) and found to be in compliance with these requirements. In addition, a copy of the CAS Final Report which was reviewed by the Administration for the issue of the Statement of Compliance and a copy of the relevant Review Record shall be placed on board to accompany the Statement of Compliance.

TITLE:	Subdivision and stability information	74
REFERENCE:	MARPOL Annex I, regulation 28	

Every oil tanker to which regulation 28 of Annex I of MARPOL applies shall be provided in an approved form with information relative to loading and distribution of cargo necessary to ensure compliance with the provisions of this regulation and data on the ability of the ship to comply with damage stability criteria as determined by this regulation.

TITLE:	STS Operation Plan and Records of STS Operations	75
REFERENCE:	MARPOL Annex I, regulation 41	

Any oil tanker involved in STS operations shall carry on board a plan prescribing how to conduct STS operations (STS operations Plan) not later than the date of the first annual, intermediate or renewal survey of the ship to be carried out on or after 1 January 2011. Each oil tanker's STS operations plan shall be approved by the Administration. The STS operations plan shall be written in the working language of the ship.

Records of STS operations shall be retained on board for three years and be readily available for inspection.

TITLE:	VOC Management Plan	76
REFERENCE:	MARPOL Annex VI, regulation 15.6	

A tanker carrying crude oil, to which MARPOL Annex VI, regulation 15.1 applies, shall have on board and implement a VOC Management Plan.

4: In addition to the certificates listed in sections 1 and 3 above, where appropriate, any ship carrying noxious liquid chemical substances in bulk shall carry:

TITLE:	International Pollution Prevention Certificate for the Carriage of Noxious Liquid Substances in Bulk (NLS Certificate)	76
REFERENCE:	MARPOL Annex II, regulation 8	

An international pollution prevention certificate for the carriage of noxious liquid substances in bulk (NLS Certificate) shall be issued, after survey in accordance with the provisions of regulation 8 of Annex II of MARPOL, to any ship carrying noxious liquid substances in bulk and which is engaged in voyages to ports or terminals under the jurisdiction of other Parties to MARPOL. In respect of chemical tankers, the Certificate of Fitness for the Carriage of Dangerous Chemicals in Bulk and the International Certificate of Fitness for the Carriage of Dangerous Chemicals in Bulk, issued under the provisions of the Bulk Chemical Code and International Bulk Chemical Code, respectively, shall have the same force and receive the same recognition as the NLS Certificate.

| TITLE: | Cargo record book | 77 |
| REFERENCE: | MARPOL Annex II, regulation 15.2 | |

Ships carrying noxious liquid substances in bulk shall be provided with a Cargo Record Book, whether as part of the ship's official log book or otherwise, in the form specified in appendix II to Annex II.

| TITLE: | Procedures and Arrangements Manual (P & A Manual) | 78 |
| REFERENCE: | MARPOL Annex II, regulation 14; resolution MEPC.18(22) | |

Every ship certified to carry noxious liquid substances in bulk shall have on board a Procedures and Arrangements Manual approved by the Administration.

| TITLE: | Shipboard Marine Pollution Emergency Plan for Noxious Liquid Substances | 76 |
| REFERENCE: | MARPOL Annex II, regulation 17 | |

Every ship of 150 gross tonnage and above certified to carry noxious liquid substances in bulk shall carry on board a shipboard marine pollution emergency plan for noxious liquid substances approved by the Administration.

5: In addition to the certificates listed in sections 1 and 3 above, where applicable, any chemical tanker shall carry:

| TITLE: | Certificate of Fitness for the Carriage of Dangerous Chemicals in Bulk | 77 |
| REFERENCE: | BCH Code, section 1.6; BCH Code, as modified by resolution MSC.18 (58), section 1.6 | |

A certificate called a Certificate of Fitness for the Carriage of Dangerous Chemicals in Bulk, the model form of which is set out in the appendix to the Bulk Chemical Code, should be issued after an initial or periodical survey to a chemical tanker engaged in international voyages which complies with the relevant requirements of the Code.

Note: The Code is mandatory under Annex II of MARPOL for chemical tankers constructed before 1 July 1986.

| | | |

| TITLE: | International Certificate of Fitness for the Carriage of Dangerous Chemicals in Bulk | 78 |
| REFERENCE: | IBC Code, section 1.5; IBC Code as modified by resolutions MSC.16(58) and MEPC.40 (29), section 1.5 | |

A certificate called an International Certificate of Fitness for the Carriage of Dangerous Chemicals in Bulk, the model form of which is set out in the appendix to the International Bulk Chemical Code, should be issued after an initial or periodical survey to a chemical tanker engaged in international voyages, which complies with the relevant requirements of the Code.

Note: The Code is mandatory under both chapter VII of SOLAS 1974 and Annex II of MARPOL for chemical tankers constructed on or after 1 July 1986.

6: In addition to the certificates listed in sections 1 and 3 above, where applicable, any gas carrier shall carry:

TITLE:	Certificate of Fitness for the Carriage of Liquefied Gases in Bulk	79
REFERENCE:	GC Code, section 1.6	

A certificate called a Certificate of Fitness for the Carriage of Liquefied Gases in Bulk, the model form of which is set out in the appendix to the Gas Carrier Code, should be issued after an initial or periodical survey to a gas carrier which complies with the relevant requirements of the Code.

TITLE:	International Certificate of Fitness for the Carriage of Liquefied Gases in Bulk	80
REFERENCE:	IGC Code, section 1.5; IGC Code, as modified by resolution MSC.17 (58), section 1.5	

A certificate called an International Certificate of Fitness for the Carriage of Liquefied Gases in Bulk, the model form of which is set out in the appendix to the International Gas Carrier Code, should be issued after an initial or periodical survey to a gas carrier which complies with the relevant requirements of the Code.

Note: The Code is mandatory under chapter VII of SOLAS 1974 for gas carriers constructed on or after 1 July 1986.

7: In addition to the certificates listed in sections 1, and 2 or 3 above, where applicable, any high-speed craft shall carry:

TITLE:	High-Speed Craft Safety Certificate	81
REFERENCE:	SOLAS 1974, regulation X/3; 1994 HSC Code, section 1.8; 2000 HSC Code, section 1.8	

A certificate called a High-Speed Craft Safety Certificate shall be issued after completion of an initial or renewal survey to a craft which complies with the requirements of the 1994 HSC Code or the 2000 HSC Code, as appropriate.

TITLE:	Permit to Operate High-Speed Craft	82
REFERENCE:	1994 HSC Code, section 1.9; 2000 HSC Code, section 1.9	

A certificate called a Permit to Operate High-Speed Craft shall be issued to a craft which complies with the requirements set out in paragraphs 1.2.2 to 1.2.7 of the 1994 HSC Code or the 2000 HSC Code, as appropriate.

8: In addition to the certificates listed in sections 1, and 2 or 3 above, where applicable, any ship carrying dangerous goods shall carry:

TITLE:	Document of compliance with the special requirements for ships carrying dangerous goods	83
REFERENCE:	SOLAS 1974, regulation II-2/19.4	

The Administration shall provide the ship with an appropriate document as evidence of compliance of construction and equipment with the requirements of regulation II-2/19 of SOLAS 1974. Certification for dangerous goods, except solid dangerous goods in bulk, is not required for those cargoes specified as class 6.2 and 7 and dangerous goods in limited quantities.

9: In addition to the certificates listed in sections 1, and 2 or 3 above, where applicable, any ship carrying dangerous goods in packaged form shall carry:		
TITLE:	**Dangerous goods manifest or stowage plan**	84
REFERENCE:	**SOLAS 1974, regulations VII/4.5 and VII/7-2; MARPOL Annex III, Regulation 4**	

Each ship carrying dangerous goods in packaged form shall have a special list or manifest setting forth, in accordance with the classification set out in the IMDG Code, the dangerous goods on board and the location thereof. Each ship carrying dangerous goods in solid form in bulk shall have a list or manifest setting forth the dangerous goods on board and the location thereof. A detailed stowage plan, which identifies by class and sets out the location of all dangerous goods on board, may be used in place of such a special list or manifest. A copy of one of these documents shall be made available before departure to the person or organization designated by the port State authority.

10: In addition to the certificates listed in sections 1, and 2 or 3 above, where applicable, any ship carrying INF cargo shall carry:		
TITLE:	**International Certificate of Fitness for the Carriage of INF Cargo**	85
REFERENCE:	**SOLAS 1974, regulation VII/16; INF Code (resolution MSC.88(71)), Paragraph 1.3**	

A ship carrying INF cargo shall comply with the requirements of the International Code for the Safe Carriage of Packaged Irradiated Nuclear Fuel, Plutonium and High-Level Radioactive Wastes on Board Ships (INF Code) in addition to any other applicable requirements of the SOLAS regulations and shall be surveyed and be provided with the International Certificate of Fitness for the Carriage of INF Cargo.

11: In addition to the certificates listed in sections 1, and 2 or 3 above, where applicable, any Nuclear Ship shall carry:		
TITLE:	**A Nuclear Cargo Ship Safety Certificate or Nuclear Passenger Ship Safety Certificate, in place of the Cargo Ship Safety Certificate or Passenger Ship Safety Certificate, as appropriate.**	86
REFERENCE:	**SOLAS 1974, regulation VIII/10**	

Every Nuclear powered ship shall be issued with the certificate required by SOLAS chapter VIII.

Certificates and other documents which are *not mandatory*:

Title: *Special Purpose Ship Safety Certificate*

In addition to SOLAS certificates as specified in paragraph 7 of the Preamble of the Code of Safety for Special Purpose Ships, a Special Purpose Ship Safety Certificate should be issued after survey in accordance with the provisions of paragraph 1.6 of the Code for Special Purpose Ships. The duration and validity of the certificate should be governed by the respective provisions for cargo ships in SOLAS 1974. If a certificate is issued for a special purpose ship of less than 500 gross tonnage, this certificate should indicate to what extent relaxations in accordance with 1.2 were accepted.

Reference: Resolution A.534 (13), as amended by MSC/Circ.739; 2008 SPS Code (resolution MSC.266 (84)), SOLAS 1974, regulation I/12; 1988 SOLAS Protocol, regulation I/12

Title: *Offshore Supply Vessel Document of Compliance*

The Document of Compliance should be issued after the vessel is found to comply with the provisions of the Guidelines for the design and construction of Offshore Supply Vessels, 2006.

Reference: Resolution MSC.235 (82)

Title: *Certificate of Fitness for Offshore Support Vessels*
When carrying such cargoes, offshore support vessels should carry a Certificate of Fitness issued under the "Guidelines for the Transport and Handling of Limited Amounts of Hazardous and Noxious Liquid Substances in Bulk on Offshore Support Vessels." If an offshore support vessel carries only noxious liquid substances, a suitably endorsed International Pollution Prevention Certificate for the Carriage of Noxious Liquid Substances in Bulk may be issued instead of the above Certificate of Fitness.
Reference: Resolution A.673 (16); MARPOL Annex II, regulation 13 (4)

Title: *Diving System Safety Certificate*
A certificate should be issued either by the Administration or any person or organization duly authorized by it after survey or inspection to a diving system which complies with the requirements of the Code of Safety for Diving Systems. In every case, the Administration should assume full responsibility for the certificate.
Reference: Resolution A.536 (13), section 1.6

Title: *Safety Compliance Certificate for Passenger Submersible Craft*
Applicable to submersible craft adapted to accommodate passengers and intended for underwater excursions with the pressure in the passenger compartment at or near one atmosphere.
A Design and Construction Document issued by the Administration should be attached to the Safety Compliance Certificate.
Reference: MSC/Circ.981, as amended by MSC/Circ.1125

Title: *Dynamically Supported Craft Construction and Equipment Certificate*
To be issued after a survey carried out in accordance with paragraph 1.5.1 (a) of the Code of Safety for Dynamically Supported Craft.
Reference: Resolution A.373 (X), section 1.6

Title: *Mobile Offshore Drilling Unit Safety Certificate*
To be issued after a survey carried out in accordance with the provisions of the Code for the Construction and Equipment of Mobile Offshore Drilling Units, 1979, or, for units constructed on or after 1 May 1991, the Code for the Construction and Equipment of Drilling Units, 1989.
Reference: Resolution A.414 (XI), section 1.6; resolution A.649 (16), section 1.6; resolution A.649 (16), as modified by resolution MSC.38 (63), section 1.6; 2009 MODU Code (resolution A.1023 (26))

Title: *Wing-in-Ground Craft Safety Certificate*
A certificate called a WIG Craft Safety Certificate should be issued after completion of an initial or renewal survey to a craft, which complies with the provisions of the Interim Guidelines for WIG craft.
Reference: MSC/Circ.1054, section 9

Title: *Permit to Operate WIG Craft*
A permit to operate should be issued by the Administration to certify compliance with the provisions of the Interim Guidelines for WIG craft.
Reference: MSC/Circ.1054, section 10

Title: *Noise Survey Report*
Applicable to existing ships to which SOLAS II-1/3-12 does not apply.
A noise survey report should be made for each ship in accordance with the Code on Noise Levels on Board Ships.
Reference: Resolution A.468 (XII), section 4.3

ON-HIRE AND OFF-HIRE CONDITION SURVEYS

GENTIME—THE DRY CARGO TIME CHARTER PARTY

The introduction gave us a brief idea of what is involved with on-hire/off-hire surveys. Here, we shall see in more detail what is involved with such surveys.

In late 1995, BIMCO set up a subcommittee whose purpose was to compose a new time charter party.

BIMCO's documentary committee had its last meeting in 1998 upon which the final draft of the General Time Charter Party was unveiled. The finalized form of this document appeared in 1999.

Clause 5—On/off-hire surveys.

Clause 5 of the "GENTIME—The Dry Cargo Time Charter Party" deals with on-hire and off-hire surveys.

This Clause provides that the on-hire survey shall be conducted without loss of time to the charterers, whereas the off-hire survey shall be conducted in the charterers' time. The wording "without loss of time" in lines 109/110 has been used in recognition that the vessel may come on-hire for instance on arrival at the pilot station at the first port of call, whereas the on-hire survey shall not take place until it arrives in the port.

It would therefore not be entirely appropriate to stipulate that the on-hire survey shall be conducted in the owners' time if the vessel is rendering the services required at the time, typically loading, as long as the on-hire survey does not hamper the loading operations.

The Time Charter Party states (lines 107-115) (sample copy here: www.bimco.org/~/media/Chartering/Document_Samples/Time_Charter_Parties/Sample_Copy_GENTIME.ashx).

Quote

> Joint on-hire and off-hire surveys shall be conducted by mutually acceptable surveyors at ports or places to be agreed. The on-hire survey shall be conducted without loss of time to the Charterers, whereas the off-hire survey shall be conducted in the Charterers' time. Surveys fees and expenses shall be shared equally between the Owners and the Charterers.
>
> Both surveys shall cover the condition of the Vessel and her equipment as well as quantities of fuels remaining onboard. The Owners shall instruct the Master to co-operate with the surveyors in conducting such surveys.

Unquote

> Clause 6—Bunkers.
> Clause 6 of the "GENTIME—The Dry Cargo Time Charter Party" deals with bunkers:
> Subclause (a) provides that the quantities of fuel at delivery and redelivery should be about the same, subject to the vessel on redelivery having enough fuel to safely reach the nearest port where bunkers of the required type or better are available.

Sub-clause (b) contains an additional provision allowing the charterers and the owners to bunker prior to delivery and redelivery, respectively. In both cases, the bunkering should be done without hindrance to the other party's operation of the vessel.

Considering the problems associated with bunkering, it was felt appropriate to provide extensive guidelines for the bunkering operation in subclause (d).

Furthermore, specific reference has been made to sampling procedures bearing in mind the corresponding provisions of BIMCO's FUELCON Standard Marine Fuels Purchasing Contract. Although the marine fuel purchasing contract is a matter between the time charterers and the fuel suppliers, the vessel's crew and, in particular, the chief engineer have an important role to play. By actively assisting the charterers in their dealings with the suppliers, they will act in the best interest of the owners. The Clause attempts to address this fact by requiring the chief engineer to co-operate with the charterers' bunkering agents and fuel suppliers.

Sub-clause (e) places full liability on the charterers for any loss or damage to the owners caused by the supply of unsuitable fuel and exonerates the owners for any reduction in performance or any other consequences arising as a result thereof. Nevertheless, the burden rests with the owners to prove that unsuitable fuel was the proximate cause of loss or damage; hence, the importance of proper sampling procedures.

The Time Charter Party states (lines 116-151):

Quote

(a) *Quantity at Delivery/Redelivery*—The Vessel shall be delivered with about the quantity of fuels stated in box 19, and unless indicated to the contrary in box 20, the Vessel shall be redelivered with about the same quantity, provided that the quantity of fuels at redelivery is at least sufficient to allow the Vessel to safely reach the nearest port at which fuels of the required type or better are available.

(b) *Bunkering prior to the Delivery and Redelivery*—Provided that it can be accomplished at scheduled ports, without hindrance, to the operation of the Vessel, and by prior arrangement between the parties, the Owners shall allow the Charterers to bunker for the account of the Charterers prior to delivery and the Charterers shall allow the Owners to bunker for the account of Owners prior to redelivery.

(c) *Purchased Price*—The Charterers shall purchase the fuels onboard at delivery at the price stated in box 21 and the Owners shall purchase the fuels on board at redelivery at the price stated in box 22. The value of the fuel on delivery shall be paid together with the first installment of hire.

(d) *Bunkering*—The Charterers shall supply fuel of the specifications and grades stated in box 23. The fuels shall be of a stable and homogeneous nature and unless otherwise agreed in writing, shall comply with ISO standard 8217:1966 or any subsequent amendments thereof as well as with the relevant provisions of Marpol. The Chief-Engineer shall co-operate with the Charterers' bunkering agents and fuel suppliers and comply with their requirements during bunkering, including but not limited to checking, verifying, and acknowledging sampling, readings, or soundings, meters, etc., before, during, and/or after delivery of fuels. During the delivery of representative samples of all fuels shall be taken at a point as close as possible to the Vessel's bunker manifold. The samples shall be labeled and sealed and signed by suppliers, Chief-Engineer, and the Charterers or their agents. Two samples shall be retained by the suppliers and one each by the Vessel and the Charterers. If any claim should arise in respect of the quality or specification or grades of the fuels supplied, the samples of the fuels retained as aforesaid shall be analyzed by a qualified and independent laboratory.

(e) *Liability*—The Charterers shall be liable for any loss or damage to the Owners caused by the supply of unsuitable fuels or fuels which do not comply with the specifications and grades set out in box 23 and the Owners shall not be held liable for any reduction in the Vessel's speed performance and/or increased bunker consumption nor for any time lost and any other consequences arising as a result of such supply.

Unquote

The foregoing have been quoted with the kind permission of BIMCO.

Similar to GENTIME the NYPE charter party states that

"Prior to delivery and redelivery the parties shall each appoint surveyors, for their respective accounts, who shall conduct joint on-hire/off-hire surveys. A single report shall be prepared on each occasion and signed by each surveyor, without prejudice to his right to file a separate report setting forth the items upon which the surveyors cannot agree.

If either party fails to have a representative attend the survey and sign the joint survey report, such party shall nevertheless be bound for the purposes by the findings in any report prepared by the other party. On-hire survey shall be at Charterers' time and off-hire survey at an Owners' time."

Both GENTIME and NYPE charter parties are similar. When it comes to on-hire and off-hire surveys, as we saw earlier, GENTIME states that both surveys shall cover the condition of the Vessel and her equipment as well as quantities of fuels remaining onboard.

It is of interest to note that there are no particular stipulations as to the parts of the ship which require to be inspected by the surveyors. It is expected that this is so because the parts of the ship which may suffer damage, during the currency of a charter, are those associated with the loading, discharging and general manipulation of cargo. In addition, the recording of the ship's certificates is also important because it proves that the Owners have kept the ship in Class. This, in turn, reflects good management.

ON-HIRE/OFF-HIRE CONDITION SURVEYS: AN OVERVIEW

A. *On-hire* and *Off-hire* surveys are arranged for the purpose of inspecting a ship at the time of her delivery to a new charter and at the time she is redelivered from a charter, respectively. These inspections provide the means which enable the attending surveyor to compile a report that contains an accurate and detailed description of

(01.00) the ship's weather decks and respective structural parts

(02.00) the ship's cargo gear

(03.00) the ship's outfit and equipment located on the weather decks

(04.00) the ship's hatch-coamings, hatch covers and associated equipment, including container fittings and securing gear

(05.00) the ship's Fo'c'sle, including all equipment and mooring gear and bossun's store

(06.00) the ship's poop deck, including all equipment and mooring gear

(07.00) in the event that the ship is outfitted with tween decks these will be inspected as to the structural members and equipment

(08.00)   the ship's lower cargo holds

(09.00)   upon inspection of the inner surfaces of the coamings, the tween decks and the lower cargo holds, the surveyor will grade the cleanliness of these parts and spaces

(010.00) on having completed the inspection of parts 1-9 above, the surveyor will proceed to inspect the hull, port and starboard, and for'd and aft (as far as visible) and will record any damage to the shell plating and the associated structural members, such as frames, etc. At the same time, the surveyor will record the condition of the ship's coatings/paintwork

(011.00) subsequently to the foregoing the surveyor will examine the ship's certificates, including Class certificates

(012.00) damage suffered by the vessel during loading and/or unloading operations, and other damage reports issued by the ship's command to the charterers and their servants (e.g., Stevedores)

(013.00) depending on the instructions received from the owners and/or the charterers the surveyor may extend his survey to the following areas:

(13.01)   navigating bridge and equipment

(13.02)   life saving appliances and equipment

(13.03)   accommodation and ancillary areas

(13.04)   engine-room and other machinery areas (e.g., steering gear)

(13.05)   scrutinizing of the "Planned Maintenance" records

B. *On-hire* and *off-hire* bunker surveys are arranged for the purpose of ascertaining the quantities of fuels remaining onboard at the time of the ship's delivery to charterers, as well as at the time of her redelivery to her owners.

(14.00) these surveys involve the sounding of all tanks containing heavy fuel oil and marine diesel oil, including service and settling tanks. Subsequent to the sounding of all of the tanks the surveyor will calculate the fuels remaining onboard at the time of his attendance, utilizing the ship's calibration tables and manuals providing the necessary correction factors, issued by recognized organizations, such as *ASTM/API/IP*.

Depending on instructions received the surveyor may have to adjust the quantities of fuels found onboard at the time of the survey to the quantities that would be remaining onboard at the time the vessel passed the point of delivery, or redelivery, in accordance with the provisions of the charter party.

It is important that the surveyor ensures that his/her principals have enumerated and described their exact requirements because these details are important in estimating the time it will take the surveyor to complete his/her attendance and most importantly the fees he/she must invoice.

In certain instances, it will be required of the surveyors carrying out an "Off-hire" survey to compare their findings to those described within the "On-hire" report, in order to establish whether any damage has been suffered by the vessel during the period of the charter party.

PERSONAL SAFETY

Before we look at the personal safety of surveyors while onboard a ship, it is important that all those who work onboard her are sufficiently protected from falling victims of an accident.

Suitable equipment must exist on a ship as well as training programs, which will protect the safety of all working onboard. Accidents will occur on a ship no matter what prevention procedures and safety equipment are available and in place. The probability of accidents is always high (whether these occur due to human errors or due to wear and tear of materials left to exist unnoticed), crew and others working on the vessel must be alert to all possible dangers.

Training, on a regular basis, as well as holding drills over regular intervals are the important steps to be taken so that all serving on a ship are constantly reminded of the possible dangers.

One of the most common causes of accidents onboard a ship is entering enclosed spaces. The main aspects to be considered with respect to each enclosed space are

(1) difficulty in entering and exiting
(2) restrictions while carrying out work
(3) planned rescue procedures in cases of emergencies

Those who have had sufficient service-time onboard a ship will be aware of all these difficulties and restrictions, however, once they are no longer onboard that knowledge disappears.

Today the *Enclosed Space Management System* attempts to provide an assessment of the risks involved with the entering of enclosed spaces onboard each ship.

Before starting his inspection the surveyor must meet with the ship's Master and officers, to discuss the safety aspects to be observed during the survey. Specifically, communication channels and rescue procedures must be established.

Hatches, or access manholes of any spaces that require ventilation must have been opened up well before the surveyor is to enter such spaces (at least 30/60 min before entry time).

If safety so requires, forced ventilation will be applied. In the event that spaces such as double bottom tanks, duct keels, cofferdams, and pipe tunnels are part of the spaces required by principals to be inspected, the surveyor must ensure that two manholes per compartment are opened so that forced ventilation can be successfully applied, well in advance of entry.

If there is any doubt as to the "fitness to enter" certain compartments their atmosphere should be tested with equipment that has recently been calibrated so that the latter provides accurate test results. If necessary, breathing apparatus must be worn.

Appropriate rescue equipment should be available at a strategically central location, from where it can be quickly deployed.

SURVEYOR'S EQUIPMENT AND TOOLS

Surveyors ought to possess a comprehensive tool kit and should carry a sufficient number of tools with them so that they can devote most of their time on inspections rather than trying to find substitutes for missing tools.

Over a period of time, a surveyor gets to know exactly what he should take with him depending on the type of survey he will be undertaking. The suggested tools and equipment is as shown in Tables 1 and 2. However, depending on the type of surveys that are undertaken on a regular basis, additional equipment such as an infrared camera and grp testing meter may be part of a surveyor's tools:

Table 1 Personal Equipment—Protection and Safety

Boiler suit
Safety boots
Hard hat
Gloves
Life jacket and personal locating beacon
Hi-visibility jacket
Protective, safety spectacles
Pair of corded foam ear plugs
Harness (scaffolds) kit
Spotlight and/or LED torch
Measuring tape
Notebook (10×15 cm), pen, and highlighter
Monocular
Digital camera (with WiFi) and macro lens
Pocket knife

Table 2 Personal Equipment—Productivity

Lightweight/portable laptop, loaded with appropriate software, forms, and certificates
USB (for handing over to Master reports and recommendations)
Petroleum measurement tables
Sounding tape
Water finding and fuel finding paste
Chalk and markers
Test hammer (brass head)
Infrared digital distant thermometer
Sample bags
Draft survey hydrometer and measuring cylinder for same
Note recording equipment
Pocket calculator
Silver nitrate test kit
Copy book (21×29 cm)
Large backpack/rucksack, for carrying most of the items in Tables 1 and 2

A surveyor's toolkit may also depend on local weather conditions. Protection against rain and cold weather is one scenario, whereas hot and humid climates present different requirements.

Finally, it is important to ensure that all necessary official passes for port entry and for boarding ships are available and up to date.

SURVEYS IN DRYDOCK

It is unusual for a ship to be inspected in drydock on account of a charter party dispute, however, it does happen that a ship must be inspected in drydock if/when the owners allege damage to have been sustained during the currency of a particular charter party and state that this damage is attributable to causes held covered by it.

In circumstances such as these the owners will choose a suitable time and place to have their ship drydocked (most likely coinciding with the anniversary of Class requirements/due dates). This drydocking will be for the account of the owners, but they will invite for joint inspections with the charterers. If the in-drydock inspections prove, beyond dispute, that the alleged damage occurred during the currency of the said charter party, then repairs would be effected.

Adjusters would be appointed to apportion the drydocking costs which will be for the owners' account and those costs payable by the charterers. These apportionments would be based on the report-findings of the surveyors attending in drydock.

When a ship enters drydock (or floating dock, or slipway), she is placed on blocks that support her hull weight and allow examination of her bottom shell. Staging will be arranged where necessary, so that the side-shell, bow and stern shell, and stern-frame and rudder can be accurately inspected (Figures 48–52).

Sometimes the surveyor may be requested to attend an "in-water" survey instead of a drydocking. In such instances a meeting will take place so that all attending are aware of how the divers inspecting the ship's bottom will proceed with their work.

Depending on the findings of this survey, it may be decided that an inspection in drydock is required.

The drydock is flooded and then its gates are opened. The ship is guided in the drydock so as to be placed exactly over the blocks, which have been arranged in accordance with the drydocking plan provided by the ship's owners. The dock gates are then closed and the water is pumped out. Once the dock is dry work commences and this covers the following aspects:

The photo on the left depicts the exposed double bottom of an OBO, after removal of the damaged frames and shell plating, in the way of the area that had been affected by the vessel having grounded on a rock face.

Bob cats, forklifts and dock cranes have been utilized to remove discarded materials and to maneuver in place new steel.

FIGURE 48

Drydock inspection of the damaged double bottom of an OBO.

In this photo new longitudinal frames and transverses have been inserted in the positions of the damaged strength members of the hull and have been welded as required.

Note the staging and lighting that have been provided to improve working conditions.

FIGURE 49

Picture shows longitudinal and transverse double-bottom framing to have been replaced with new frames.

Pad eyes have been welded to the shell plating to assist with the lifting of the new materials used in the repairs.

FIGURE 50

Picture shows new longitudinals being welded in place at the after end of the affected double bottom area.

(1) *Condition of the shell plating*

The outside of the shell plating, including the bilge keels, will be visually inspected from stern to stem, on both port and starboard sides. The points to be recorded are

(1a) mechanical damage, including indents and other distortions

(1b) wear due to corrosion

(1c) cracks

(1d) marine growth, accumulation

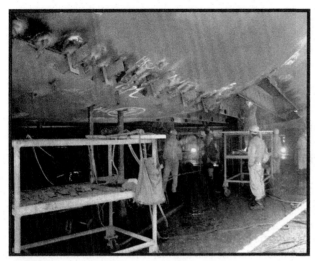

The new shell plating has been provided with a curved surface, in the yard's steel workshop, that exactly fits the original shape of the ship's shell.

As can be seen the shell plating has been temporarily welded and wedged in order to make the final adjustments for properly fitting it to the internals, before final welding takes place.

FIGURE 51

Picture of the new shell plating having been put in place, at the after end of the affected double bottom area.

All plating, in the way of the completed repairs, has been fully ground to remove any unwanted protrusions.

On completion of the welding works the plating will be washed with a high pressure water jet for the purpose of cleaning it and making it ready for painting with the required undercoats and top coat.

FIGURE 52

Picture of the completed repairs with the new outer-shell plating having been fully welded in place. Paint undercoats have been applied and allowed to properly dry prior to applying a topcoat.

Points (1a)-(1c) may lead to shell plating renewals and even internals' replacement. This work, depending on the nature and extent of damage, may require the attendance of a Class surveyor.

All of the above points will necessitate the cleaning of the shell plating prior to any other work commencing.

Cleaning can be carried out by

(1e) scraping and wire-brushing

(1f) high pressure water jets, which can remove all growths and paint, so as to expose the base metal

(1g) shot blasting with abrasive

(1h) a combination of (1f) and (1g)

Once the cleaning has been effected any damaged shell plating and internals are removed and renewed with materials of the same specification.

Subsequently the outer-shell plating is first coated with anticorrosive paint. Parts of the shell plating that remain submerged are painted with a coat of antifouling paint. The number of coats of anticorrosive and antifouling to be applied will depend on owners' requirements.

(2) *Condition of the tailshaft*

Measurements are taken to establish the clearance between the sterntube bearings and the tailshaft. Depending on the findings the tailshaft may have to be removed for further inspection. This happens after removal of the propeller.

The tailshaft will be cleaned and then magnetic-particle tested so as to ascertain whether any cracks exist. The shaft's wear is also recorded and may be condemned if the wear is found to be excessive. Otherwise, the shaft is polished.

(3) *Condition of the sterntube*

The sterntube becomes fully exposed on removal of the tailshaft. At that time, the sterntube bearings are measured, and the clearance between tailshaft and bearings are established. Bearings are either made of lignum vitae and are water cooled, or they are made of white metal and they are cooled by lubricating oil.

If the clearances are found to be excessive, then the existing lignum vitae staves will be removed and replaced by new staves of larger size (so as to reduce the clearance to the required size). Equally, if the clearances are found to be excessive, then the existing white metal will be built up (so as to reduce the clearance between the bearings and the tailshaft to the required size).

(4) *Condition of the sterngland*

With the ship in drydock it is a good opportunity to dismantle the sterngland, clean it, and pack it with new material.

(5) *Condition of the propeller*

The propeller, while in situ, can be inspected for cavitation effects, corrosion, and damage to its tips. As we saw in (2) above, the propeller is removed from the tailshaft so as to allow the latter's removal.

Cavitation effects can be repaired by building up the material where necessary and grinding it so that the blade surface is even and smooth.

Any damage to the propeller tips can be repaired in a similar manner. Depending on the format of repairs, annealing may be required.

(6) *Condition of the rudder*

A ship's rudder is constructed by the production of a framework consisting of aerofoil shaped horizontal sections, which are held in position by edges located at the forward and after end of the overall rudder construction.

Shell plating is placed so as to cover all of the foil sections, on the port and starboard sides of the rudder (Figure 53).

FIGURE 53

Photo shows the lower part of the rudder having been lost at sea due to the propagation of cracks.

The aft part of the rudder moves to an angle that is greater than the angle of the main part of it (nearly twice as great).

FIGURE 54

The photo shows a "flap-rudder."

Source: Rolls Royce.

Rudders can suffer failures due to corrosion, cracks, and deformations. So all of these aspects will require attention during a survey in drydock, including taking and appraising the clearances of the pintles (Figure 54).

There are many different types of rudders fitted on different ships. The type of rudder depends on the shape of a ship's stern and on the capacity of the individual ship's steering gear.

However, there are two types of rudders which are widely used (Figures 55 and 56).

(1) Semi-balanced skeg rudder.
(2) Balanced rudder.

In recent years other types of rudder have been developed to suit particular needs. Schottel produce a so called "RudderPropeller" which incorporates a propeller in a cylindrical shroud. This combination produces maneuverability and efficiency.

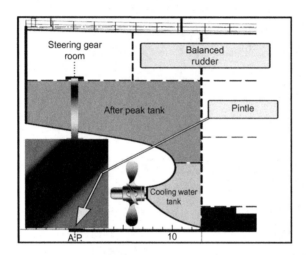

FIGURE 55

Balanced rudder.

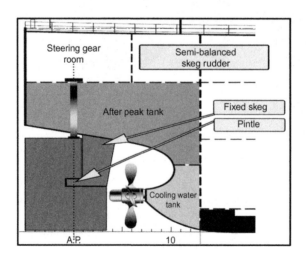

FIGURE 56

Semi-balanced skeg rudder.

Voith-Schneider produce a combination of a propulsion unit and a rudder with the capability of almost instantaneous thrust-change which results in a rudder action. This type of equipment is best suited for harbor-tugs and ferry boats.

CLASS SURVEYS IN DRYDOCK

A ship will undergo a *Class Renewal Survey*, CRS (or *Special Survey*, SS) every 5 years, or at a maximum interval of 63 months, provided that the Classification Society so agree. The owners may be allowed to commence the "CRS" on the fourth "Annual Survey" with a programmed inspection that leads to the completion of all necessary surveys on the anniversary of the fifth "Annual Survey".

The survey of the external part of the shell plating, including other related equipment will take place in drydock. Accordingly, it is called a *Docking Survey*. In lieu of a *Docking Survey* the ship, at the discretion of her Classification Society, may be examined while afloat by way of an *In-water Survey*.

While the *Class Renewal Survey* is held every 5 years the two *Docking Surveys* (or one *Docking Survey* and one *In-water Survey*) must be held over a maximum interval period of 36 months. One of these two surveys must be held at the same time with the *Class Renewal Survey* (Figures 57 and 58).

Large drydocks usually accommodate more than one ship at a time.

This is in order for the shipyard to maximize its earnings (or to speed up repayments of the craving dock building costs).

In this photo three ships are docked concurrently, one of them is a large tanker.

FIGURE 57

Ships in drydock. Note the ranged anchor cables and blocks.

(a) During the ship's "Class Renewal Survey" in drydock the propeller shafts and the intermediate shafts will be visually inspected, measured and (usually) magnetic-particle tested.

(b) The stern bearing clearances will be taken and after removal of the shafts the bearing will be visually examined for any undue localized wearing.

(c) The clearances of the rudder bearings will also be taken.

Once (a)-(c) have been tested and the findings have been recorded a memorandum will be prepared and copies will be handed over to the Classification Society's surveyor and to the owners' representative.

FIGURE 58

View of a ship in drydock. The ship's stern, hull, and rudder have been cleaned and repainted. The propeller has been cleaned and polished.

CASE FILE MANAGEMENT AND COMMENCEMENT OF SURVEY

RECEIPT OF A SURVEY-APPLICATION

A firm of Marine Surveyors will receive an application from the Owners of a ship, or from her Charterers, requesting attendance for a condition survey, with a view to assessing her condition. Such surveys are held either at the time of the commencement of a Charter (On-Hire Condition Survey), or at the time of completion of the Charter (Off-Hire Condition Survey).

All those employed within an office of Marine Surveyors, must be thoroughly familiar with how to react upon receiving a work-application.

The secretariat must quickly react in accordance with the procedures established within their office, for receiving survey-requests, or related inquiries and for dealing with them in the utmost professional manner. A senior member of the office must be appropriately consulted.

The "Request for Attendance" is received from the "Applicant," either by telephone, e-mail, or fax. In the event that the "Request for Attendance" is received by phone, the "Surveyors" will request the "Applicant" to fax or e-mail the same, together with any related documents, which are required for proceeding with the execution of the assignment.

The "Surveyors" will scrutinize all of the incoming documents to ensure that there is a clear description of what the "Applicant" requires. If the information received is not complete (or sufficient) the "Surveyors" must go back to the "Applicant" and ask for any additional information that they may require.

There are occasions when the ship condition survey may have to commence with the ship in dry dock and complete with the ship afloat. If this is the case the surveyor must be aware of what parts of the ship are to be examined while she is in dry dock and then he must suitably arrange to attend to the balance of his inspection when the ship is afloat.

Having looked at this possibility the surveyor ought to ascertain—as best as circumstances permit—the ship's itinerary once she leaves dry dock. In the event that the ship is to depart to a foreign port, arrangements will have to be made with her Owners and the Charterers for her to remain at a suitable berth or at anchorage for sufficient time so as to complete the inspection of all of the items required to be surveyed in accordance with the Owners' and the Charterers' instructions.

It is important that when receiving instructions for a survey, the surveyor should be clear in his mind as to whom he is acting for and what exactly is required of him under his appointment.

The details of any verbal communication will be clearly documented in the "Case Notes" and in "Field Attendance Record."

As soon as time permits the salient points of any verbal communications/exchanges will be sent to the "Applicants" by e-mail or fax.

CONFIRMING SURVEY ATTENDANCE

The surveyors must respond to a work-application advising the applicants:

(1) That their requirements are clearly understood.
(2) That they have the required knowledge and expertise to complete the work satisfactorily.

An indication of the currency and the amount that will be charged for the work requested must be included in the initial response.

It is extremely important that on sending their initial response the surveying firm's *Terms and Conditions of Business* are attached for the perusal of the applicants.

The "Surveyors" will ensure that the local agents of the vessel have been copied in of their appointment by the "Applicant."

PLANNING AND PREPARING FOR THE SURVEY

The next step will be for the "Surveyors" to prepare their "Case Notes" and "Field Attendance" record, or whatever is the usual documentation that forms part of their case-file. The "Surveyors" will choose the person that has the capability to meet all of the work-assignment's requirements and they will forward his/her name to the "Applicant" and the local agent accordingly.

The "Surveyors" will ensure that the case-file contains copies of

1 "Case Notes and Field Attendance" record.
2 "Inspection report" (blank copy).
3 "Bunkers' report" (blank copy).
4 "Bunkers' certificates" (blank copy).
5 "Urgent Recommendations" report (blank copy).
6 Other documents and/or manuals which the appointed surveyor may require.

At this time the "Surveyors" may request a deposit, if this is considered appropriate.

The "Surveyors" will respond to any correspondence received from an "Applicant" (whether such correspondence may be an inquiry or a firm work order) within 1 hour of its receipt, or on opening the next working day (if it is received after closing).

Correspondence of a general nature, which does not specifically pertain to existing work, or to a firm order, may be replied to within 24 hours of its receipt.

The survey should be conducted within two working days (or as soon as circumstances permit, or as soon as it is practical) after receiving instructions.

Where the payment is required in advance, in part, or in full, then the survey may commence within two working days after receipt of funds. If this is not possible, agree a date of survey with the "Applicant" and notify all parties involved.

A request for an urgent survey should be attended to immediately. If the survey, for well-founded reasons, cannot be conducted there and then, contact the "Applicant" to discuss and agree the most appropriate course of action.

The "Surveyor" firm will seek to ensure that the following information is available as early as possible:

(a) The "Applicant's" contact details (name of the person, office and mobile numbers, location of survey).
(b) Ship's name and arrival and departure details.

(c) Time the survey is required to commence (day, night, during cargo operations, etc.)

(d) Place of the survey.

(e) Persons to be advised of the survey, such as agents, ship's staff, other surveyors (in case of joint survey), including agreed time of boarding/survey.

(f) Survey schedule/sequence to propose (crew meal breaks, coffee times, time between shifts).

(g) Whether their surveyor is the sole surveyor, or whether a joint survey is to be held and, if so, to identify who the other surveyors are and whom they represent.

The surveyors shall use all reasonable care and skill in the performance of their services in accordance with sound marine surveying practice.

The surveyors will not disclose any information provided to them in confidence by the "Applicant" to any third party and they will not permit access to such information by any third party, unless the "Applicant" expressly grants permission, save where it may be required to do so by an order of a competent court of law. All the documents received from the "Applicant" are to be placed in the "Work Assignment" file.

In due course the formal report of the attending surveyor will be forwarded to the "Applicant" in accordance with any agreement made when the case-file was opened, however, if the surveyor has any urgent recommendations he should hand these over to the ship's Master and he will also send these to the "Applicant" by e-mail or fax.

COMMENCEMENT OF SURVEY

Irrespectively as to whether the surveyor is acting independently, or jointly with another surveyor who is acting for another party, the following process is recommended:

(1) The surveyor will board the vessel and present himself to the ship's Master to whom he/she will explain the purpose of this attendance.

(2) Discuss any ongoing cargo operations with the Master and agree a survey procedure that will not interfere with the ship's operations or unduly delay the vessel.

(3) The surveyor must seek to learn from the Master whether any enclosed spaces, subject to inspection, are safe for entry.

(4) All safety steps to be taken will be discussed and agreed with the Master of the vessel, as well as with the ship's officer who will accompany the surveyor.

(5) If a bunker survey is to be held at the same time as the ship condition survey this is a good time to establish together with the sip's command which of these two surveys should be conducted first.

(6) In the event that the ship is to be inspected while in dry dock the surveyor will have already been advised by the Owners, or by the Charterers as to which are the items to be inspected. The surveyor knows that his attendance is not associated with Class requirements and procedures; accordingly he/she must follow a rotation, which is closer to a survey required by Hull and Machinery underwriters. In all probability, either the Owners or the Charterers (or both) may have already notified their H&M underwriters and they will—in due course—submit the survey-findings and their possible claim to them.

There are instances in which the surveyor is acting directly for the H&M underwriters. If this is the case, then the following (notes no 1-22 below) ought to be taken into account.

(1) The shipowners utilize the services of a broker to negotiate the Terms and Conditions of a Marine Insurance Policy with the Underwriters.

(2) The Underwriters will negotiate the final settlement with the Shipowner, on the basis of all the claim documentation.

(3) There are exceptions, but generally speaking, the Surveyor is not required to know policy terms and conditions and should carefully avoid discussion of same. The underwriters should notify the surveyor of any particular terms that he needs to observe, or be aware of.

(4) The attending Surveyor should note the damage and comment, within his report, on whether he considers it a result of the incident(s) claimed. Under no circumstances the surveyor is to commit the Underwriters, or express any opinions to any third parties, including the owners' representative.

(5) The surveyor will hold the survey in accordance with his capabilities which are based on qualification and experience. If there are matters that fall outside these he ought to contact the underwriters who may decide to provide him with additional expertise, particularly if there is a conflict of opinion on the cause of damage, or its extent.

(6) The reason a survey is held is so that the surveyor ascertains the cause, nature, and extent of damage. The surveyor will keep the underwriters advised with a preliminary advice after the first survey, this to be followed by intermediate reports. Finally, the surveyor will prepare a comprehensive report, with all appropriate documents, drawings, and photos attached.

(7) If necessary—before repairs are put in hand—the surveyor and the owners' representative should consult the surveyor acting for the ship's classification society as any repairs must be in conformity of the classification society's requirements, provided that these do not go beyond what may be termed a reasonable repair to the damage. In cases such as this, the surveyor will notify the underwriters immediately and make a suitable entry in his survey report.

(8) The owners will instruct their representative to present the parts that are being claimed as part of the damage to the surveyor so that the latter can record these and agree that they are part of the required repairs.

(9) The accounts of the repairs should be presented by the owners to the surveyor for his perusal. As soon as the surveyor has had the opportunity to check the accounts he should go over them with the owners' representative and point out any of those that he may not agree with. An appointment should be made with the repairers so that the accounts can be negotiated. Any overtime costs should be compared to the time saved afloat and in drydock. Accounts submitted to the surveyor covering repairs which he has not sighted should not be signed, but he should notify the underwriters.

(10) It is not unusual for owners to request their underwriters to make interim payments or payments on account. All such accounts must be approved by the surveyor before any payment is made. The surveyor is required to provide an opinion as to the reasonable cost of the repairs necessary to rectify the alleged damage.

(11) Any items considered to be outside, the scope of the damage claimed may be approved under the notation that these are subject to the surveyor's comments in his report. A separate section within the report must clearly show such items together with their corresponding costs and the underwriters' attention should be drawn to them.

(12) The ship's log-books must be requested and scrutinized via certified translations, to a language he understands, over a period the surveyor considers sufficient in order to draw accurate conclusions. If the surveyor understands the language in which the log-books are written he should make sure that they do not appear to have been altered/amended in any way.

(13) Classification certificates, or any other documentation considered relevant may be requested from the owners. If the information that has been requested has not been provided a note should be inserted in the surveyor's report.

(14) Discussions regarding the claimed repairs particulars, between the surveyor and the owners' representative, may lead to either disagreement with regard to a number of items, or to doubt. If there is disagreement, or doubt on the part of the surveyor, he should advise the owners' representative accordingly, but he may accept that further evidence may be submitted by the owners in support of their claim. If such evidence is not forthcoming for some time a note should be entered in the surveyor's report, or in his advices to underwriters.

(15) Soon after the surveyor has completed the initial survey he should advise the underwriters of
   (a) the damage claimed
   (b) the repairs claimed
   (c) the allegation of the cause of damage
   (d) the comments he has, thus far, and of the likely cost of repairs
   (e) any joint field survey reports he has signed (with copies of such documents attached)
      Note: These field survey reports ought to solely cover the description of the damage found. If there is any mention of the cause, or alleged cause of the casualty, the surveyor should not sign these documents. Any items which, in the surveyor's opinion are not related to the casualty should be endorsed as such.
   (f) an estimate of the amount and time of drydocking—if held without carrying out repairs
   (g) an estimate of the drydock services
   (h) an estimate of tug-hire
   (i) an estimate of the total costs of repairs
   (j) an estimate of the cost of repairs—drydocking costs excluded
   (k) an estimate of the cost of any deferred repairs

(16) The surveyor's final (formal) report must be written in such a manner that it can be fully understood by all those who will have access to it, e.g., underwriters, lawyers, average adjusters, owners' claims' department.

(17) The final report ought to be written and forwarded to the party that applied for it soon after the repairs have been completed. In the event that any information is missing at the time the final report is ready this should not delay its forwarding to the applicants. Once the missing information is made available the surveyor should scrutinize all of it and compile an addendum, with all of the relevant documentation attached, and send this to the applicants as quickly as possible.

(18) The surveyor must make a statement in his report regarding the cause of damage alleged by the owners' representative. He should also make a statement as to whether he agrees, disagrees or has any reservations/doubts as to the alleged cause of damage.

(19) If the owners provide no allegation on the cause of the damage, then no statement in this connection should be included in the surveyor's report covering his own opinion regarding the cause of damage. If this is the case the surveyor should send a separate statement to the applicants, providing his opinion on the cause of damage. Along similar lines the surveyor may declare that he requires additional information before he can make an emphatic statement of his opinion relative to the cause of damage. The details of the required information must be stated.

(20) The surveyor, within his formal report, ought to provide a tabulated account of the repairs, shown as follows.

Description of Damage Found	Damage Related to the Alleged Casualty	Repairs Recommended	Repairs Carried Out	Cost of Repairs (US$)	Reasonable Cost of Repairs (US$)	Remarks
Description-1	Yes	Description-1A	Description-1A	123	123	-
Description-2	No	-	Description-2A	237	200	Com-1
Description-3	Yes	Description-3A	Description-4A	456	348	Com-2
Description-4	Yes	Description-6A	Repairs deferred	-	-	Com-3

Row Description 1
- The damage has been attributed by the surveyor to the alleged casualty.
- The repairs carried out are the same as the repairs recommended by the surveyor.
- The cost of the repairs is equal to the cost the surveyor considers to be reasonable.
- No remarks have been entered as there are none required.

Row Description 2
- The damage has not been attributed by the surveyor to the alleged casualty.
- Because the damage, according to the surveyor, is not connected with the alleged casualty there is no description required under "Repairs Recommended."
- Description-2A is given purely for guidance/reference.
- Cost of repairs is given purely for guidance/reference.
- Comments Com-1 have been entered to show why this item is not connected with the alleged casualty.

Row Description 3
- The damage has been attributed by the surveyor to the alleged casualty.
- The repairs carried out are not the same as the repairs recommended by the surveyor.
- The cost of the repairs is not equal to the cost the surveyor considers to be reasonable.
- Comments Com-2 have been entered to show why this item is not connected with the alleged casualty and to justify the amount given by the surveyor as "Reasonable Cost of Repairs."

Row Description 4
- The damage has been attributed by the surveyor to the alleged casualty.
- The repairs recommended by the surveyor have been described for later reference.
- The repairs have been deferred and this has been shown under "Repairs Carried Out."
- The cost of the repairs is omitted as no repairs have been carried out this time.
- Comments Com-3 have been entered to show why this item is connected with the alleged casualty. Formal reports should not contain cost estimates. These should be forwarded to the applicants under separate cover.

Note: The Surveyor, within his concluding remarks, must declare that (a) the issuing of his report, as well as (b) the inclusion in his report of the noted costs of the repairs is so given *without prejudice to liability.*

(21) Owners' repairs may have been carried out at the same time as damage repairs. In such cases the Surveyor should state in his report whether the owner's work involved drydocking and whether the owner's work was strictly necessary for the safety of the vessel and its cargo on the ensuing voyage. The time taken in respect of work afloat or in drydock, for the owners' account should be shown.

(22) There are a number of other areas in which the surveyor may be called upon, by the party that has appointed him/her, to provide comments and advice.

- Work and costs involved with cleaning of cargo tanks.
- Description of any work considered as supplementary, including comments on expenditure.
- Specification of shell painting and prices charged.
- Replacement of damaged parts with new parts—new for old.
- Details and costs of temporary repairs.
- Work executed, which is considered as "improvements."
- Comments on various superimposed damages.
- Repairs or replacements of parts subject to wear and tear.
- Apportionment of credits for parts discarded.
- Repairs held which are not related to a particular casualty, but are for owner's work.
- Detailed breakdown of the period of time under repair.
- Identification of Machinery subject to damage related to a casualty.
- Particulars of surveys held without prejudice.
- Comments regarding responsibility and damage specification related to collisions.
- Collision appraisals, including angle of blow and speed.
- Circumstances and causes giving rise to loss of hire/earnings.
- Damage arising out of crew negligence.
- Commenting with respect of damage for which the cause is unknown.
- Investigations concerning reported fuel shortage.
- The causes leading to the ship putting into a port of refuge.
- Repairs on parts damaged by latent defects.
- Description and expenditure covering salvage operations.
- Liability investigations and surveys of damage to docks, piers, and other fixed objects.
- Findings on the condition of ship's anchors and cables.
- Investigations covering charterer's legal liability.
- Surveys of stevedore damage.
- Ship surveys for cargo interests.
- Surveys and accounts' appraisals regarding General Average.
- Surveys held covering voyage and towage approval, including commenting on submitted costs.
- Scrutinizing the circumstances which may have led to conflict(s) of interest.

SEQUENCE OF INSPECTION OF SHIP PARTS - PART 1

A comparison between a general dry-cargo ship and a bulk carrier will serve well to offer an introduction to this section. The classification of ships depends on the trades they are built for. So let's look at what these trades are as well as what different types of ships can accommodate. Generally there are three main groups of ships, as follows:

At first glance bulk carriers started their illustrious working lives as the ships of a specialized design that enabled them to carry bulk grain, cement, various kinds of ore and coal. After years of operations (spent exclusively carrying bulk cargoes) the market returns for this type of ship declined. The pressures exerted on their owners resulted in the latter experimenting with the carriage of general cargoes.

Those owners who exercised careful and meticulous cargo stowing, introduced to the shipping markets the bulk carrier that could successfully carry general cargoes. The cargoes carried ranged from bagged rice and fertilizer, timber products, steel reinforcing bars, and project cargoes such as mobile homes and prefabricated frame-units of different shapes and sizes.

Not only was a new life invented for the bulk carrier, but because of superior carrying capacity the bulk carrier in this role could compete with general cargo ships engaged in similar trades. Especially the geared bulk carriers, which could handle cargo loading and unloading in ports without cargo handling cranes.

Even the overall construction of bulk carriers and general cargo ships converged in the sense that soon both of these types of ships had their machinery spaces and accommodation arranged aft, with all of the cargo holds arranged forward of the accommodation bulkhead. The different categories of cargo ships may be summarized as follows:

Group 1	General Cargo Carriers	Tween Deckers	
		Bulk Carriers	
		Other Ships	Container ships.
			Reefer ships.
			Heavy Lift ships, etc.
Group 2	Passenger Ships	Limited Range services	
		Transatlantic Cruiseliners	
Group 3	Specialized Purpose Ships	Trawlers	Local services, deep sea.
		Tugs, etc.	Harbor,ocean going.

The major design difference between these ships lay in the fact that the general cargo ships continued to utilize tween decks, which are capable of stowing on the sides of the hold so as to create an uninterrupted cargo space. This space, then extended from the top of the tank top to the undersides of the hatch covers.

Even a comparison of container ships (with the exception of the cargoes that are carried on such a type of ship) can show that their outline configuration is not entirely dissimilar to that of a tween deck general cargo carrier and/or a bulk carrier.

A note of caution here—today's Triple E class super container ships have their accommodation arranged approximately one-third from forward, with a superstructure arranged over the engine-room and accommodating the ship's funnel, at a distance of nearly one-third from aft. This is shown in Figure 59.

FIGURE 59

New generation container ship "CAP SAN LORENZO," 9814 TEUs, 2100 reefer plugs.

Source: Hamburg Süd.

Having arrived at this point in our pondering the sequence of inspection of the various ship parts and areas becomes a lot clearer than when we were about to start our comparisons.

So putting aside the particular design of a ship we can start examining the sequence in which a surveyor would go about carrying out his ship-condition inspection.

Figure 60 shows two side views and the plan view of a four-hold bulk carrier. We shall use this diagram to plot the route that is best for the inspection of this vessel.

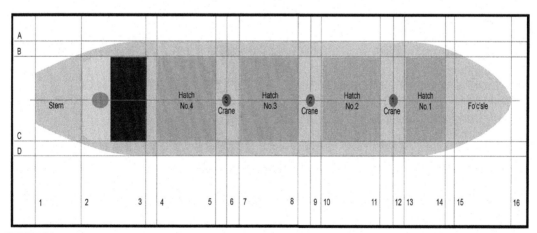

FIGURE 60

A diagrammatic plan view of a four-hold bulk carrier.

The surveyor will start his condition survey by accessing the weather-deck from the poop-deck.

(1) Weather deck and associated equipment in area C3-D3, D3-D4, D4-C4, and C4-C3.
Inspect:
the bulwark (or railings)
ventilators
fairleads
multipurpose chocks
warping rollers
double bollards
sounding pipes

(2) Cross-deck and associated equipment in area C3-B3, B3-B4, B4-C4, and C4-C3.
Inspect:
accommodation forward bulkhead
all attached sounding pipes
ventilators
doors to accommodation
electrical fittings and lights
fire-hose boxes and their contents
after transverse coaming and its support stiffeners/stays
piping and electrical cable trays/piping
hatch-cover hydraulic pistons and associated equipment, alternatively, hatch-cover rollers
and roller-assemblies
quick acting cleats, cross-joint cleats, and wedges
eccentric wheel assemblies
balancing roller assemblies
junction pieces
top cleat assemblies
kenter shackles
nonreturn valves
ramp stoppers

(3) Weather deck and associated equipment in area C4-C5, C5-D5, D5-D4, and D4-C4.
Inspect:
the bulwark (or railings)
ventilators
fairleads
multipurpose chocks
warping rollers
double bollards
sounding pipes
timber posts (if any)
quick acting cleats, cross-joint cleats, and wedges
eccentric wheel assemblies
balancing roller assemblies

junction pieces
top cleat assemblies
kenter shackles
nonreturn valves
ramp stoppers
stbd longitudinal coaming and its support stiffeners/stays
all vents and fire-dampers and fire extinguishing lines attached to the hatch coamings.
Repeat for all hatch coamings on the stbd-side
piping and electrical cable trays/piping
steam and/or compressed-air lines

(4) Weather deck and associated equipment in area C5-C7, C7-D7, D7-D5, and D5-C5.
Inspect:
the bulwark (or railings)
ventilators
fairleads
multipurpose chocks
warping rollers
double bollards
sounding pipes
timber posts (if any)
crossover from weather-deck side to cross-deck side
for'd transverse coaming and its support stiffeners/stays
piping and electrical cable trays/piping
hatch-cover hydraulic pistons and associated equipment, alternatively, hatch-cover rollers
and roller-assemblies
quick acting cleats, cross-joint cleats, and wedges
eccentric wheel assemblies
nonreturn valves
access hatches to cargo holds
inspect integrity of the crane pedestal. If space internal to the pedestal is accessible through
a deck-door confirm its proper function as well as the efficient condition of its rubber seal
witness the proper working of the crane controls while the latter is operated by a crew member

(5) Inspect the rest of the weather-deck areas, hatch covers, hatch coamings, and cranes as
described in (1)-(4) previously.
access the bossun's store, for'd of hatch no. 1 and inspect hawse pipes, chain wash piping
and related valves. Confirm whether a spare anchor is mounted on the fo'c'sle bulkhead.
Fo'c'sle bulkhead
all attached sounding pipes
ventilators
doors to bossun's store
electrical fittings and lights
fire-hose boxes and their contents

(6) Inspect access ladders to fo'c'sle.

(7) Inspect mast and its equipment.

(8) Inspect anchor windlass and anchor stoppers.
   ventilators
   fairleads
   multipurpose chocks
   warping rollers
   double bollards
   sounding pipes

(9) Repeat the process described between (1) and (6) on the port side weather deck, including any steam and/or compressed-air lines.

(10) Once these inspections have been completed, inspect the hatch-cover tops, including cross-joint wedges and container fittings. In direct continuation ask for each hatch cover (or for each set of hatch covers) to be opened in your presence. Observe the even opening of the covers and once satisfied that these operate satisfactorily proceed with the inspection of
   hatch-coaming flats
   compression bars
   inner surfaces of the hatch coamings, including hatch corner curvatures
   balancing rollers
   eccentric rollers
   quick acting cleats
   drainage system
   nonreturn drainage valves
   parts of the covers that land on the coaming-flats (steel-to-steel contact)
   rubber gaskets
   if so instructed carry out a hose test, or an ultrasonic test of the hatch covers

(11) Access the cargo holds, via the deck hatches. Once on the tank top maintain the same inspection-rotation for all the cargo holds. By doing this you will better remember where your findings and photos correspond.
   Starting at the after port corner, hold your inspection moving clockwise.
   Inspect:
   Port-side
   the slopping plating of the lower hopper tanks
   the side shell frames, including their webs and upper and lower brackets (look out for cracks and for grooving)
   the side shell
   manholes in the tank top and on the slopping plating of the lower hoppers
   any and all vertical piping, including guard arrangements for the same
   For'd-bulkhead
   lower stool slopping plating
   lower shelf plate, shedder plate
   corrugated transverse bulkhead plating
   ends of upper shelf plate
   cross deck structure cantilever support structure

cargo hatchway end transverse beam

ladders and various piping, within corrugations

Stbd-side

identical to the Port-side

After-bulkhead

identical to the For'd-bulkhead

Tank top

condition of plating and of any manholes (mainly located in the after end transverse area), including bilge boxes, strum boxes, and ends of sounding pipes

if in any doubt ask for the bilge suctions to be tested

Mechanical damage

Any mechanical damage is identified as follows:

Size: Length, width, depth.

Location: Distance from the center of the damaged area to a prominent edge of a port, or stbd-side (whichever is nearer), distance of the center of the damaged area from a prominent edge on the for'd, or the after bulkhead (whichever is nearer).

Similarly, if the damage is located on the slopping plating of the hopper tanks, or bulkheads, its distance should be measured from an easily identifiable reference point.

Designation:

Slight indent (depth up to 1.50 cm).

Moderate indent (depth up to 2.50 cm).

Heavy indent (depth over 2.50 cm).

(12) On completion of the inspection of the cargo holds (including inspection of the tween decks, if the survey concerns a general cargo tween decker) the surveyor ought to continue his/her work with the inspection of any other areas that have been requested by the applicants. Alternatively, he/she may commence the bunker survey.

The above constitute a suggested sequence of inspecting the various parts of the deck, cranes, and cargo holds. The surveyor will always choose the best process based on the type of ship involved and on the requirements of the applicants.

SEQUENCE OF INSPECTION OF SHIP PARTS - PART 2

On holding an *On/Off-Hire Condition Survey* it is suggested that the sequence of inspection of the various ship parts and equipment be arranged as shown in the following tables. The lists presented concern the inspection of a four-hold bulk-carrier and they can easily be adopted to comply with the actual requirements of the party applying for the survey. Figures 61–66 shown below are for quick reference purposes.

The suggested sequence of inspection of the various ship parts and equipment covers a four-hold bulk carrier. This type of ship has been chosen because of construction and equipment similarities with general cargo carriers and other similar dry cargo vessels.

Ship parts should be inspected with a view to establishing

(1) the condition of the paint coating

(2) the mechanical condition of the constituent members (e.g., cracks, indents, and fractures)

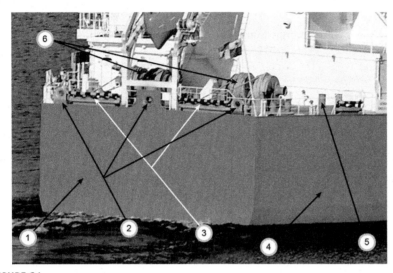

1: Transom Stern
2: Panama Lead
3: Roller Chocks
4: Stbd Quarter Shell Plating
5: Mooring Bitts
6: Stern Winches

FIGURE 61

Stern and stern-equipment. Bulk carrier.

Photo: Hamburg Süd.

1: Navigating Bridge Deck
2: Bridge wing
3: Flying Bridge
4: Captain's Accom. Deck
5: Officers' Accom.
6: Accom. access doors
7: Stbd navigation light
8: Boat deck
9: Accom. External ladders

FIGURE 62

Accommodation, stbd-side. Bulk carrier.

Photo: Hamburg Süd.

1: Inflatable liferaft
2: Accom. & E.R. ventilators
3: Accom. Portholes
4: Semi rigid raft
5: After stbd draft marks
6: Stbd gangway

FIGURE 63

Accommodation and stbd quarter. Bulk carrier.

Photo: Hamburg Süd.

1: Hatch covers
2: Hatch cover ventilation cover
3: Hatch coamings
4: Weather deck
5: Pilot ladder

FIGURE 64

Weather deck and hatch coamings/covers. Bulk carrier.

Photo: Hamburg Süd.

1: Observation/service platform
2: Crane boom
3: Operator's cab
4: Revolving superstructure
5: Boom securing crutch
6: Cargo grab
7: Pedestal deck-house
8: Crane hook and pulley block
9: Crane top with wire sheaves
10: Boom heel pin

FIGURE 65

Weather deck and hatch coamings/covers. Cargo gear. Bulk carrier.

1: Anchor and anchor-housing
2: Stbd bow
3: Rails
4: Door to bossun's stores
5: Acces ladder to the fo'c'sle
6: Apron platform and pulpit
7: Stbd windlass with mooring drum storage spool and warping drums
8: Fo'c'sle mast
9: Platform at the mast's top, incl. foghorn
10: Bulwark
11: Guide rollers and multipurpose chocks/fairleads

FIGURE 66

Fo'c'sle equipment and outfit. Bulk carrier.

Photo: Hamburg Süd.

(**3**) the standard of upkeep and maintenance

(**4**) the necessity to take immediate remedial action due to the unacceptably unsafe condition of a particular item

	Remarks on Findings			
1. Hull	1	2	3	4
Port bow side shell plating				
Port side shell plating				
Port side stern side shell plating				
Transom stern side shell plating				
Stbd side stern side shell plating				
Stbd side shell plating				
Stbd side bow shell plating				
Port side boot-topping				
Stbd side boot-topping				
Bulbous bow				
Port anchor (or cable, if anchor down)				
Stbd anchor (or cable, if anchor down)				
Port side aft Draft marks				
Stbd side aft Draft marks				
Port side amidships draft marks				
Stbd side amidships draft marks				
Port side Loadline marks				
Stbd side Loadline marks				
Port side For'd draft marks				
Stbd side For'd draft marks				
Gangway				
Anchors (P&S)				

It will be necessary to employ a launch in order to inspect the side of the ship opposite to the one on which she is moored/berthed.

	Remarks on Findings			
2. Stern Decks	1	2	3	4
Port side aft deck shell plating, equipment and accommodation				
Stern deck shell plating, equipment and accommodation and mooring winches				
Stbd side aft deck shell plating, equipment and accommodation				

	Remarks on Findings			
3. Weather Deck-Port	1	2	3	4
Port side weather deck shell plating				
Port side bulwark and railings				

	Remarks on Findings			
3. Weather Deck-Port	1	2	3	4
Port side coamings and hatch covers Hatch 1 and Fo'c'sle Blkhd Cross deck structure, outfit and equipment Hatch 1 and Hatch 2 Cross deck structure, outfit and equipment Hatch 2 and Hatch 3 Cross deck structure, outfit and equipment Hatch 3 and Hatch 4 Cross deck structure, outfit and equipment Hatch 4 and Accommodation Blkhd Cross deck structure, outfit and equipment				

	Remarks on Findings			
4. Fo'c'sle Deck	1	2	3	4
Fo'c'sle deck shell plating, equipment/outfit Windlass and assoc. equipment Mooring winches Bossun's stores Chain locker Fore Peak tank (internal) Chain locker Emergency fire pump				

	Remarks on Findings			
5. Weather Deck-Stbd	1	2	3	4
Stbd side weather deck shell plating Stbd side bulwark and railings Stbd side coamings and hatch covers				

	Remarks on Findings			
6. Deck Cranes and Crane Pedestals	1	2	3	4
Crane No.1 (For'd), pedestal and pedestal house Crane No.2, pedestal and pedestal house Crane No.3, pedestal and pedestal house				

7. Topside Tanks-Port Internal Inspection	Remarks on Findings			
	1	2	3	4
No.1 Topside Port tank (For'd)				
No.2 Topside Port tank				
No.3 Topside Port tank				
No.4 Topside Port tank				

8. Topside Tanks-Stbd Internal Inspection	Remarks on Findings			
	1	2	3	4
No.1 Topside Stbd tank (For'd)				
No.2 Topside Stbd tank				
No.3 Topside Stbd tank				
No.4 Topside Stbd tank				

9. Cargo Holds-Hold No.1	Remarks on Findings			
	1	2	3	4
Port side shell plating, frames, hopper and topside tanks, attached equipment				
For'd bulkhead shell plating, corrugations, attached equipment				
Stbd side shell plating, frames, hopper and topside tanks, attached equipment				
After bulkhead shell plating, corrugations, attached equipment				
Tanktop including bilge suctions (P&S)				

10. Tanks-Hold No.1		Remarks on Findings			
		1	2	3	4
Port side WB topside tank	Visual				
Port side WB hopper tank	Visual/Internal				
Stbd side WB topside tank	Visual				
Stbd side WB hopper tank	Visual/Internal				
Port side DB, WB tank	Internal				
Stbd side DB, WB tank	Internal				
Center DB, WB tank	Internal				

11. Cargo Holds-Hold No.2	Remarks on Findings			
	1	2	3	4
Port side shell plating, frames, hopper and topside tanks, attached equipment				
For'd bulkhead shell plating, corrugations, attached equipment				

	Remarks on Findings			
11. Cargo Holds-Hold No.2	1	2	3	4
Stbd side shell plating, frames, hopper and topside tanks, attached equipment				
After bulkhead shell plating, corrugations, attached equipment				
Tanktop including bilge suctions (P&S)				

		Remarks on Findings			
12. Tanks-Hold No.2		1	2	3	4
Port side WB topside tank	Visual				
Port side WB hopper tank	Visual/Internal				
Stbd side WB topside tank	Visual				
Stbd side WB hopper tank	Visual/Internal				
Port side DB, WB tank	Internal				
Stbd side DB, WB tank	Internal				
Center DB, WB tank	Internal				

	Remarks on Findings			
13. Cargo Holds-Hold No.3	1	2	3	4
Port side shell plating, frames, hopper and topside tanks, attached equipment				
For'd bulkhead shell plating, corrugations, attached equipment				
Stbd side shell plating, frames, hopper and topside tanks, attached equipment				
After bulkhead shell plating, corrugations, attached equipment				
Tanktop including bilge suctions (P&S)				

		Remarks on Findings			
14. Tanks-Hold No.3		1	2	3	4
Port side WB topside tank	Visual				
Port side WB hopper tank	Visual/Internal				
Stbd side WB topside tank	Visual				
Stbd side WB hopper tank	Visual/Internal				
Port side DB, WB tank	Internal				
Stbd side DB, WB tank	Internal				
Center DB, WB tank	Internal				

15. Cargo Holds-Hold No.4	Remarks on Findings			
	1	2	3	4
Port side shell plating, frames, hopper and topside tanks, attached equipment				
For'd bulkhead shell plating, corrugations, attached equipment				
Stbd side shell plating, frames, hopper and topside tanks, attached equipment				
After bulkhead shell plating, corrugations, attached equipment				
Tanktop including bilge suctions (P&S)				

16. Tanks-Hold No.4		Remarks on Findings			
		1	2	3	4
Port side WB topside tank	Visual				
Port side WB hopper tank	Visual/Internal				
Stbd side WB topside tank	Visual				
Stbd side WB hopper tank	Visual/Internal				
Port side DB, WB tank	Internal				
Stbd side DB, WB tank	Internal				
Center DB, WB tank	Internal				
Cargo holds Fixed fire extinguishing systems					
Cargo holds fire detection systems					

17. Accommodation				Remarks on Findings			
				1	2	3	4
Crew's accommodation							
Officers' accommodation							
Passengers' accommodation							
Superstructure, externally							
Superstructure external ladders							
Accommodation, internal staircases and stairs							
Galley							
Heating, Ventilation, Air-Conditioning Rooms							
Refrigerated Store Rooms							
Dairy	*Fish*	*Meat*	*Vegetables*	*Other*			
Store Rooms (1), (2)							
Portable Safety Appliances							
Fixed fire extinguishing systems							
Fire detection systems							

	Remarks on Findings			
18. Deck Equipment and Outfit	1	2	3	4
Rescue boat Lifebuoys Liferafts-Port, incl. hydrostatic releases Liferafts-Stbd, incl. hydrostatic releases Liferafts (other locations), incl. hydrostatic releases Lifeboats and Davits, Port-side Lifeboats and Davits, Stbd-side Stores' crane(s) E.R. Fuel shut-off mechanisms				

	Remarks on Findings			
19. Engine-Room Equipment and Outfit	1	2	3	4
Portable Safety Appliances Fire Extinguishing Systems E.R. Fire detection systems Remote Shut-off for an emergency generator				

	Remarks on Findings			
19. Engine-Room Equipment and Outfit	1	2	3	4
Main-Engine				
General remarks Cylinder liner wear Crankshaft deflections Spare parts				
Auxiliary Diesel Generators				
General remarks Cylinder liner wear Crankshaft deflections Spare parts				
Other Machinery				
Pumps Condensers Evaporators Compressors Fresh Water Generators Incinerators				

	Remarks on Findings			
19. Engine-Room Equipment and Outfit	1	2	3	4
Separators/Purifiers Refrigerating equipment: Dairy Refrigerating equipment: Fish Refrigerating equipment: Meat Refrigerating equipment: Vegetable Emergency fire pump *Boiler installation(s)* Fixed Fire Extinguishing system for boiler(s) *CO_2 Room*				
Tailshaft, Propeller, Rudder				
Tailshaft condition and clearances, last docking Spare tailshaft Propeller condition at last drydock Spare propeller Duct keel Rudder condition and clearances, last docking				
Steering Gear				
Steering gear and repeaters Bridge communications system Emergency steering gear				

	Remarks on Findings			
19. Engine-Room Equipment and Outfit	1	2	3	4
Main Switchboard				
Instruments Controls Internal cleanliness Wiring and Cable termination Nameplates, Warning signs				
Emergency Switchboard				
Instruments Controls Internal cleanliness Wiring and Cable termination Nameplates, Warning signs Automatic Emergency Generator starting				

19. Engine-Room Equipment and Outfit	Remarks on Findings			
	1	2	3	4
Cables				
Terminations				
Cable Tray/clips				
Identification				
General condition				
Shore connection				
Distribution boards, starters, control panels and consoles				

20. Electrical Systems	Remarks on Findings			
	1	2	3	4
Accommodation				
Deck				
Cargo holds				

21. Navigation Bridge	Remarks on Findings			
	1	2	3	4
External communications				
High Frequency/Medium Frequency Radio				
Satcom				
VHF sets				
Lifeboat VHF sets				
EPIRBs/SARTs				
Weatherfax				

21. Navigation Bridge	Remarks on Findings			
	1	2	3	4
Internal communications				
Auto telephones				
Battery/sound powered telephones				
PA system				
Walky Talky sets				
Navigation instruments				
Anemometer				
Autopilot				
Chronometers				
Echo Sounder				

21. Navigation Bridge	Remarks on Findings			
	1	2	3	4
Engine remote control				
Gyro compass				
Loading computer				
LOG				
Magnetic compass				
Navigational signal lights				
Positioning equipment				
Radars				
Rudder angle indicators				
Thruster control				
Whistles				

COATING BREAKDOWN AND CORROSION

Ballast tanks are required to be surveyed periodically for coating breakdown and corrosion. *Lloyd's Register Rules and Regulations for the Classification of Ships*, include this requirement.

Figure 67 has been reproduced with the kind permission of Lloyd's Register .

GOOD	condition with only minor spot rusting affecting not more than 20 per cent of areas under consideration, e.g. on a deck transverse, side transverse, on the total area of platings and stiffeners on the longitudinal structure between these components, etc.
FAIR	condition with local breakdown at edges of stiffeners and weld connections and/or light rusting over 20 per cent or More of areas under consideration.
POOR	condition with general breakdown of coating over 20 per cent or more of areas or hard scale at 10 per cen or more of areas under consideration.

FIGURE 67

Coating condition definitions.

Source: Lloyd's Register.

Some graphical tools have also been appended here to assist with the assessment of the coating condition.

In this connection the affected *area determination* and the *assessment scale for breakdown* are shown on the next page.

Figure 68 gives a good guideline to the process of determining the extend as a percentage of the area considered to which the plating coating has been affected (Figure 69).

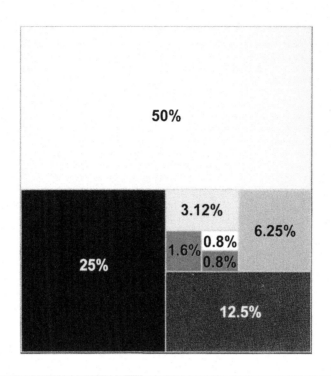

FIGURE 68

Area determination chart.

Source: Lloyd's Register.

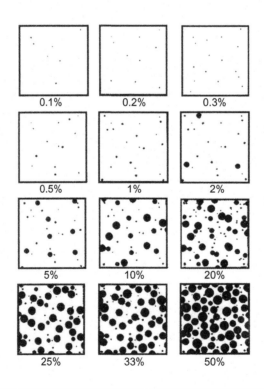

FIGURE 69

Assessment scale for breakdown.

Source: Lloyd's Register.

The definition of coating conditions is as follows (Figure 70).

Rating/condition	Good	Fair	Poor
Spot rust	Minor	>20%	
Light rust	Minor		
Edges	<20 %	>20%	
weld			
Hard scale	Minor	<10%	>10%
General breakdown	Minor	<20%	>20%

FIGURE 70

Definitions of coating condition. Note: The lowest rating within any category shall govern the final rating.

Source: Lloyd's Register.

The definitions given within Section "Coating Breakdown and Corrosion" have been used in Section "Ship Condition Survey (Part Two)" of Chapter 12, starting in p. 222, for the purpose of assessing the coating condition of the various parts of a ship and her equipment.

SHIP PARTS: GROUP 1

KEEL

DESCRIPTION

The spine of a ship is her keel, which runs from the stern to the bow. Accordingly, it contributes to her longitudinal strength, and it distributes loads created when the ship is in dry dock. Today, the majority of ships are constructed with a flat-plate keel which is internally supported by a continuous center girder. An alternative form of construction provides for two vertical side girders which are connected by a stiffener arranged on the flat keel and by a stiffener under the tank top level (Figures 71 and 72).

This arrangement forms a box structure which is known as duct keel. The duct keel allows piping of various services to extend from the engine-room to the fore-peak, intermittently extending to the double bottom tanks on the port and starboard sides. In addition, all the double bottom frames start, transversely, from the keel and extend to the port and starboard sides where they connect to the structure of the hopper tanks, or if such tanks do not exist to the margin plates (see Figures 39 and 41).

FIGURE 71

Ship's keel.

FIGURE 72

Double bottom structure. Solid/plate floors and longitudinal framing visible.

INSPECTION

The most appropriate occasion to inspect a ship's keel, for obvious reasons, is when she undergoes drydocking. However, if necessary the keel can be inspected by divers with the ship afloat.

DOUBLE BOTTOM

DESCRIPTION

A double bottom construction arrangement is an improvement to the original designs of the ships that were only outfitted with a single outer shell. Double bottoms can be used to store ballast water or fuel oil, while their inner shell was a barrier to flooding in the event the ship ran aground and the bottom outer shell was pierced/breached.

The double bottoms within a particular part of a ship, say cargo hold no. 2, may be divided longitudinally and/or transversely in order to form separate tanks. However, the wells which serve the bilges (located aft, on the port, and starboard sides of the cargo hold) are constructed to be completely

FIGURE 73

Construction of a bulk-carrier's transverse and longitudinal framing in the double bottom tanks. Galvanic sacrificial anode on frame, LHS photo.

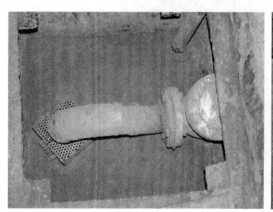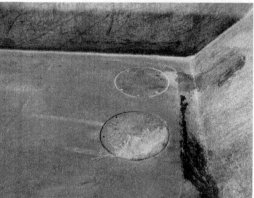

FIGURE 74

Bilge suction with strum box. Next photo shows the bilge-well cover.

independent of the double bottom spaces. As the photos in Figure 73 show the construction of the double bottom provides a robust steel member that acts together with the other ship-parts (like the decks, frames, and the bulkheads, to name but a few) toward the overall strength of the ship. The bilges in each cargo hold serve to pump out the water used in the cleaning of these spaces.

The bilge pipes are fitted with filters (strum boxes) at their ends, inside the bilge wells, in order to prevent cargo residues from entering and blocking the bilge piping. The line to the bilge pump in the E.R. is fitted with a nonreturn valve (see Figure 73) so that even if (by mistake) a pumping action toward the well is started, no water will flood the cargo hold via the bilge well. Obviously, these valves must be tested and an entry must be made in the planned maintenance record, as well as in the Log-Book. When the cargo hold is to be loaded with cargo (especially grain), a hessian cloth is used on top of the well cover (Figure 74).

Important note: Often the sounding-pipes for the bilge wells start on the weather deck and are led through the top side wing tanks before they end up in the bilge wells. It is important that the portions of the sounding-pipes, which pass through the top side wing tanks, are regularly inspected. If they develop holes (most often due to wear and tear) and the top side tanks are pumped with ballast water, this water will flow down to the bilge wells, causing flooding of the cargo hold, with consequent damage to the cargo, if the ship is in loaded condition.

INSPECTION

The inspection of the ship's double bottom and bilges can be carried out when the cargo holds are free of cargo (i.e., it is not necessary to drydock the vessel), by removing the double bottom manholes and the protective covers of the bilges, respectively. Double bottom tanks should be internally inspected for excessive corrosion, condition of protective coating (if any), and condition of the cathodic protection—if fitted and last but not least, mechanical damage (Figure 75).

FIGURE 75

Damaged keel repairs. Next photo shows new keel portions welded in place.

DESCRIPTION

Bilge keels consist of a bulb plate that is welded on a flat bar situated at the turn of the bilge, on port, and starboard sides. They extend approximately over a third of the length of the ship, with their ends tapered. The purpose of the bilge keels is to reduce the rolling of the ship.

Bilge keels are prone to damage, so they must be inspected on each dry docking and repaired if required.

Passenger ships employ active stabilizers, rather than static bilge keels.

INSPECTION

The ship's bilge keels are usually inspected when the ship is in drydock. If there is water ingress while the ship is at sea, this—most likely—will end up in the double bottom tanks; accordingly, repairs can be deferred until the ship goes to drydock. If there is serious damage that needs urgent attention, the bilge keels can be inspected and repaired by divers with the ship afloat.

BOW AND FORE END STRUCTURE/STEM, FORE PEAK TANK, AND HAWSEPIPES

DESCRIPTION

(1) *Bow and fore end structure*

The fore end of a ship is the part of its structure that extends forward of the collision bulkhead.

The construction is of such a nature as to be able to withstand panting and slamming, which are mainly encountered at the ends of a ship when she is pitching (Figure 76).

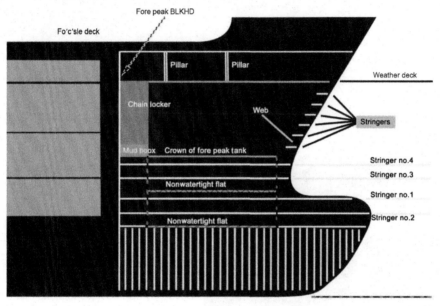

FIGURE 76

Construction details of a ship's bow.

Deep floors in the lower part of the bow in conjunction with stringers, beams, and breasthooks combine to increase strength.

Other heavy pieces of equipment, such as the windlass, are supported by increased fo'c'sle shell plating and pillars within the bossun's stores; similarly, the chain locker, in which the chains of the port and starboard anchors are stowed, is supported by the floors and the flats below it.

Figure 77 shows the anchor chains and the "bitter-end" (the chain end which is attached to the plating of the chain lockers).

(2) *Stem (see Figure 78)*

(3) *Fore peak tank*

The fore peak tank is arranged forward of the collision bulkhead (for'd bulkhead of cargo hold no. 1), and it occupies the part of the hull immediately below the chain locker and the bossun's stores. The framing may be transverse or longitudinal, and the bulbous bow forms part of the fore peak.

FIGURE 77

Anchor chains in the chain locker. Next photo shows the chain ends, attached to the locker-plating.

The photo on the left shows the way the stem of this ship has been built.
The stem is made up of plates that are shaped to follow the drawing-office bow design.

The shell plates above the bulbous bow datum have the same thickness as that of the side-shell plating.

Internally the plating is supported by webs which are arranged horizontally, center-line webs may also be fitted to make the arrangement of the stem sufficiently strong.

FIGURE 78

Plated stem and bow thruster.

A wash longitudinal bulkhead may be arranged forward of the chain locker with a solid bulkhead below it. Both of these bulkheads are arranged on the ship's centerline. The fore peak is built with adequate panting beams, extending from the port to the starboard shell plating (Figure 79).

(4) *Hawsepipes*

Hawsepipes are manufactured of cast iron and the fo'c'sle plating is increased in the way of the pipe (or a doubler can be used as an alternative). Simultaneously, the internals in the vicinity of the hawsepipes are increased in size to afford greater strength in the area that carries the weight of the anchor chains and the windlass (in many modern ships, two windlasses are fitted, one for the port and one for the stbd anchor). The diameter of the hawsepipe should be at least nine times the size of the anchor chain and the angle of the hawsepipe to the vertical should not be more than 45° (Figures 80 and 81).

(5) *Bow thruster (see Figure 82)*

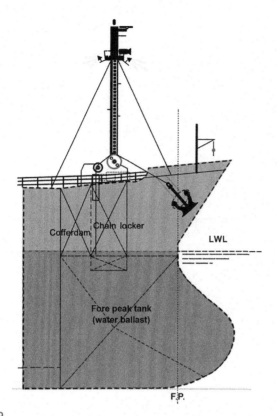

Panting stringers are also arranged on the port and starboard sides of the fore peak and at their after end they extend across the entire width of the collision bulkhead aft. Floors are incorporated within the fore peak and the bulbous bow

FIGURE 79

Ship's fore peak tank.

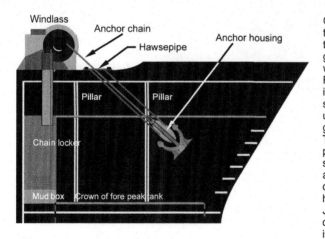

On heaving the anchor the chain is led through the hawsepipe to the gypsy-wheel of the windlass.

From there it is directed into the chain-locker via a spurling-pipe located just under the windlass' gypsy-wheel.

The hawse pipe is provided with a piping system that washes the anchor, from mud and other deposits, as it is heaved.

Just aft of the hawse pipe on the fo'c'sle deck there is a chain stopper which, when engaged does not allow any movement of the anchor chain.

FIGURE 80

The red line shows the path followed by the anchor chain as it is directed to the chain locker.

The vessel, shown in the photo on the left, is a large bulk-carrier.
As a consequence of her size she is provided with two windlasses, plus an additional warping winch, which provides additional means for mooring.
Heavy items which are ordinarily stowed in the bossun's store can be lifted to the fo'c'sle deck by means of the davit that is positioned above the access hatch.

FIGURE 81

Fo'c'sle deck, anchor windlasses, and mooring equipment.

Source: Rolls Royce.

Ships that require high maneuverability, are outfitted with bow thrusters.
Generally, only one bow thruster is all that is fitted, however, on occasions where unassisted berthing and unberthing is anticipated there may be two bow thrusters as well as two stern thrusters.
This arrangement entails the construction of ducts that can accept two propellers, one on the port and one on the starboard end of the duct.
The ducts are arranged to act transversely and do the job of pusher tugs.

FIGURE 82

Bow thruster.

Source: Rolls Royce.

INSPECTION

(1) *Bow and fore end structure*

The bow and fore end structure may be examined with the ship afloat, but depending on the reasons of the inspection scaffolding/staging may be necessary. Equally, trimming of the vessel might help with this inspection depending on which part is to be examined. Internal examination will also clarify the nature and extent of a damaged area. Damages which cannot be exposed by deballasting and trimming may have to be inspected by divers. If their nature is not hampering the ship's trading, such inspections may be postponed until the ship drydocks.

(2) *Stem*

The remarks given in "Bow and Fore End Structure" equally apply to the inspection of a ship's stem.

(3) *Hawsepipes*

The ship's hawsepipes can be inspected from the fo'c'sle, as well as by entering the bossun's stores. In the event that there is damage to the anchor pockets at the lower end of the hawsepipes, a platform may be lowered from the fo'c'sle which the crew can access using a pilot ladder.

(4) *Bow thruster*

Damage to a bow thruster, which cannot be exposed by deballasting and trimming the ship, may have to be inspected by divers. In the event that the damage cannot be repaired by divers, the class surveyor will need to be consulted as to how to proceed.

STERN AND AFTER END STRUCTURE, AFTER PEAK TANK
DESCRIPTION

(1) *Stern and after end structure*

A ship's stern requires that the naval architect designs it in such a way that the water flow from the ship's sides and bottom meets the propeller in the most efficient manner. Equally, the water should leave the propeller in a way that helps the overall propulsive effort and does not create unwanted effects to the hull and the propeller itself (Figure 83).

Today, most cargo ships are constructed with a transom stern since this type of design offers a large stern deck area. Test-tank results show that this type of stern affords improved hydrodynamic properties. Once the general outline of the stern has been decided on, further consideration will be given to

FIGURE 83

Stern/after end of a bulk carrier. Next photo shows stern/after end of a containership.

the arrangements of the propeller(s), their brackets to the hull and the rudder(s), allowing for appropriate clearances between the hull, the stern frame, and the propeller(s) in order to avoid vibration.

In addition, the stern of a ship is subject to slamming, so additional strengthening is incorporated here with the use of solid floors and vertical stiffeners. Particular attention is required where the stern frame meets the hull since it has to be of a robust construction so as to counter any vibration that may be created by the revolving propeller.

(2) *After peak tank*

The after peak tank is located in the stern area of a ship, aft of the aftermost watertight bulkhead. It serves to store fresh or ballast water. This tank, when considered in conjunction with deep tanks, the double bottom tanks, and the fore peak tank, forms a set of concentrated force-centers, which influence a ship's stability.

The after peak tank construction incorporates, the stern frame, solid floors, a centerline longitudinal wash bulkhead, a rudder trunk, and a stern tube (Figure 84).

INSPECTION

The comments given in this section equally apply to the "Stern and After End Structure." The After Peak Tank can be inspected in a manner similar to that employed in the inspection of a Fore Peak Tank.

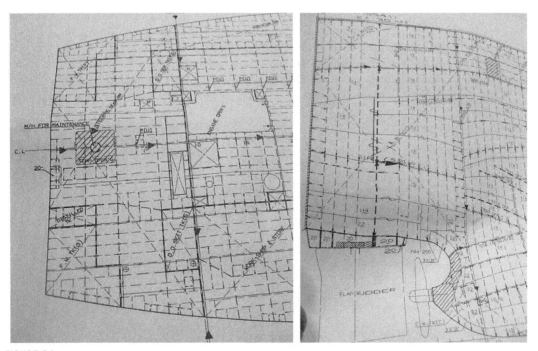

FIGURE 84

Shell expansion. Flat above the A.P. Tank (LHS) and side view of the stern arrangement (RHS).

STERN FRAME RUDDER
DESCRIPTION

(1) *Stern frame*

The importance of the stern frame lies in the fact that it supports the tailshaft and the rudder of a ship. In older ships, the stern frame was cast and then welded to the ship's outer-shell plating. It is the usual practice that the stern frame is made at a location away from the shipyard. In order to facilitate the fabrication process, the transportation to the shipyard, as well as the erection it may be required that the stern frame is made of separate parts (especially where large ships are involved). Once the stern frame is arranged in its intended position the separate parts are welded together (Figure 85).

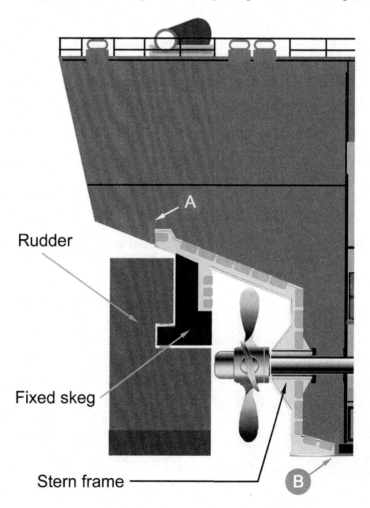

FIGURE 85

Stern frame and rudder.

Stern frames can also be fabricated from plates which are stiffened as necessary. The overall construction of a stern frame is of a smooth aerodynamic cross-section which does not produce vortices at some distance behind it.

(2) *Rudder*

Details on rudders have already been given in section "Surveys in Drydock," under part "6 Condition of the Rudder" in Chapter 4. In this section, we shall look at the maneuverability of ships, in the sense that they should be able to remain on a straight line course, or turn when necessary, without any interference due to the prevailing weather conditions.

As a set of first considerations, the designer ought to examine:

(1) *How easy will it be for a ship to maintain her directional course.*
(2) *How a ship will react when a turn begins or when she moves out of a turn and back in a straight line.*
(3) *How, or whether a ship will be able to execute a given radius full circle turn.*

When man took to the seas on primitive hand-made boats, an oar was used to make the craft change direction while in the water. Today, ocean going vessels employ a rudder in order to achieve a turn in the desired direction.

Experience and practice have shown that the most appropriate location for the position of a rudder is aft of the propeller. This is because their action is assisted by the water flow and speed produced by the propeller.

The cross-sectional area of a rudder resembles an aerofoil. In some cases, the rudder system resembles a double aerofoil, as shown in Figure 86.

Over recent years, the designer has tried to come up with a rudder that can produce a high lift with as little drag as possible. However, the shape of the ship's hull does influence the results one tries to achieve.

A short and full form gives rise to poor controls. The naval architect will specify the time it should take a ship to change direction when her rudder is turned and he/she would work toward designing a

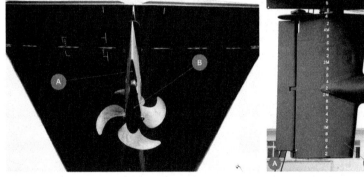

FIGURE 86

Flap rudder.

Source: Rolls Royce.

rudder that can achieve this goal, while limiting, to the minimum, the distance a ship carries on in the direction she was following before a turn order was given.

The parameters described above will be directly subject to the ship's speed, the rudder-angle, and the shape and size of the rudder itself.

Unfortunately, there are design factors which oppose each other regarding the results they produce, so at the end of the day certain compromises will have to be accepted when designing a rudder for a particular ship.

INSPECTION

(1) *Stern frame*

The inspection of the stern frame may, depending on the circumstances, be held by accessing the after peak tank. If necessary, this tank may have to be emptied for a more accurate inspection. Depending on how much the ship can be trimmed, a substantial part of the stern frame can be exposed for holding a visual inspection.

(2) *Rudder*

The rudder of a ship, to a certain extent, can be inspected by deballasting and trimming and by using a launch to access the part subject to inspection. Damage to a rudder, which cannot be exposed by deballasting and trimming, will have to be inspected by divers. In the event that the damage cannot be repaired by divers, the ship's class surveyor will need to be advised of the damage so that he can decide whether the ship can proceed to her next port of call or whether she has to enter drydock for permanent repairs.

HULL MARINE COATINGS
DESCRIPTION
What is the purpose of ship maintenance

Is ship paint-maintenance (PM) an unnecessary expense? Absolutely not. Paint maintenance is a worthwhile expense because without it ships will fall into disrepair, which brings with it a loss in value. Lack of maintenance will also lead to an overall loss of fitness for purpose.

PM consists of three stages.

(1) *Pretreatment*
(2) *Work execution*
(3) *The quality of the paint used*

The first and second stages above are more important than the third.

The main purpose of the paint maintenance is the protection of the steel from rust or corrosion.

Iron and steel are extracted from iron ore in a furnace or electro oven. The iron ore with coal or coke is heated to a very high temperature. This process introduces large amounts of energy into the ore, and eventually, this energy is stored in the iron and steel.

Oxygen is present in the furnace where the iron is produced, but when the iron leaves the furnace environment (where the oxygen is present in abundance), it becomes unstable and attempts to change to a condition similar to that which existed in the furnace.

This situation—when exposed to the atmosphere in which we live and breathe—causes the iron or its alloys to undergo an electric as well as a chemical reaction. In effect this exposure results in the gradual degradation of iron and its alloys (metals).

Chlorine and sodium are present in salt water as sodium chloride and when they unite an imbalance of electrons is generated, which, in turn, creates positive and negative ions which conduct electricity. These ions will try to move between available electrodes. The more disparate these electrodes (metals) are, the more the potential difference between them will be, as well as the strength of the current which is produced in this process.

Rust on steel surfaces takes the form of a red dust, which is what remains from the ongoing demolition of an anodic terminal.

The foregoing shows how rust comes into being, but it also provides the solution to a problem. How do we stop rust from happening? The obvious answer is to prevent the reactions described above from taking place. Oxygen in the air as well as moisture are the required agents for the rust process to start, so if we can stop air or moisture or both from coming in contact with steel, we can stop rust from happening.

We can achieve the isolation of steel surfaces from the effects described above by applying a coat of paint, but before we do so, we must ensure that the surfaces we are about to paint are free of any rust deposits because these contain oxygen and water.

Painting a rusty surface is a wasted effort and money. So, this is where the pretreatment we mentioned earlier must be put to work, and the better the pretreatment, the better the paint will serve its purpose.

So far, we looked at how rust affects exposed metals. When these metals are in sea water (like the submerged part of a ship's hull), the protection we seek will be in the form of paint coatings and to assist them to do their work better; we can also enforce a cathodic protection (in the form of sacrificial anodes or by the introduction of impressed current).

Another effect we face is fouling of the hull by barnacles and other similar sea organisms. This situation is undesirable because their presence will give rise to rust formation on the ship's hull. However, they also cause an increase in resistance which, in turn, leads to an increase in fuel consumption. This means greater than necessary operational costs for the shipowner.

Today, paint providers can supply an array of paints that will protect a ship's structure from corroding, no matter whether we are considering the superstructure, the cargo holds, or the ship's hull. We must always remember to apply the appropriate pretreatment and the correct work execution, combined with the use of a good quality paint.

SHIP PARTS: GROUP 2

7

FRAMES

DESCRIPTION

The framing shown in Figure 87 is exposed and can be easily inspected either by accessing a ship's tween deck (note the box arrangement which offers additional strength to the ship's hull structure) or in the case of a bulk carrier by reaching the tanktop. The visible framing in a bulk carrier is attached to the ship's side shell plating between the hopper tanks and the topside tanks. In the case of bulk carriers, the framing is longitudinal with wash bulkheads and web framing arranged transversely. Frames are made of bulb bars or angle bars (Figure 88).

INSPECTION

On inspecting the transverse frames, particular attention must be paid to the welding between the brackets and the plating of the topside as well as the hopper tanks. These are areas where cracks are likely to occur. Brackets that are subject to small fractures can have the damaged areas cleaned, with the end of the fracture drilled. The cleaned edges are then welded. Employing soft toe bracket ends may be the appropriate way to proceed. Large cracks may necessitate the cropping of the damaged area and the welding of an insert.

FIGURE 87

Tween-deck transverse framing. The next photo shows the transverse framing of a bulk carrier.

The termination of the framing of the side shell plating at the upper end (topside tank) and lower end (hopper tank) is by welded brackets, as shown in the photo on the LHS.

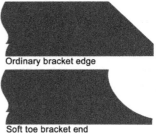

Ordinary bracket edge

Soft toe bracket end

FIGURE 88

Bracket at the foot of a transverse frame.

BEAMS
DESCRIPTION

Transverse and longitudinal beams serve two purposes: (1) they support the deck plating when it is loaded with external weights, such as cargo or heavy seas and (2) they resist lateral pressure when water pressure acts against the ship's hull. Longitudinal beams act as a supplement to the ship's overall longitudinal strength.

INSPECTION

Beams (A) connect with the frames by a bracket, called "beam-knee." This bracket is welded to both the frame and the beam and may suffer cracks where they are welded, or where the bracket meets either the beam or the frame (Figure 89).

TRANSVERSE BULKHEADS
DESCRIPTION

We have already seen some of the details on bulkheads in Chapter 3. The transverse bulkheads of a ship serve to carry the vertical loads that she encounters when she floats in calm water, when she is sailing, or when she encounters adverse weather conditions. They also provide appropriate resistance of the entire hull structure to unwanted torque. Simultaneously, the bulkheads provide the means for the ship to be subdivided into a number of watertight partitions.

FIGURE 89

Crossdeck beams.

   While this section is concerned with transverse bulkheads, we must also take into account the fact that there are occasions when longitudinal bulkheads must be arranged (especially in the case of tankers and other vessels that carry liquid cargoes).

   In designing a dry cargo vessel, the transverse bulkheads to be incorporated will depend on her length and on whether her machinery is arranged aft or at amidships. Bulkheads usually extend to the top continuous deck, but depending on freeboard assignment, there may be variations to where the bulkhead ends.

INSPECTION

The bulkheads of the fore peak tank and the after peak tank will be tested with a water head up to the load waterline. Other bulkheads are hose tested. The inspection of bulkheads is accomplished by using a cherry-picker with an articulated arm that will enable the surveyor to reach remote parts of the structure (Figure 90).

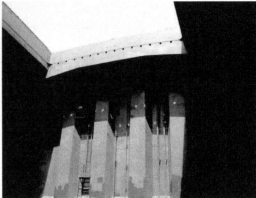

FIGURE 90

Bulk carrier vessel, corrugated bulkheads.

PILLARS

DESCRIPTION

Pillars are a means of transmitting loads from decks and other parts of a ship to her double-bottom construction, where they can be taken up by buoyancy. They can be found in the cargo holds of tween-deck ships, reefer ships, and cross-channel ferries. Pillars are also utilized in the machinery spaces of various types of ships, including bulk carriers and OBO carriers.

In instances where pillars are fitted on the center line, longitudinally, the ends of this arrangement may be outfitted with partial bulkheads that may be perforated at places. Pillars arranged in the lower hold are usually much more substantial in construction when compared with the pillars provided on the ship's tween decks. In certain designs, pillars are employed in the lower hold, whereas partial longitudinal bulkheads are fitted at the ends of the tween deck (Figures 91 and 92).

FIGURE 91

Tween-deck vessel on the LHS with pillar in the lower hold and partial bulkhead in the tween deck. The RHS photo shows the partial bulkhead (B) in the tween deck.

FIGURE 92

Pillars supporting a walkway between tween hatch covers, I tween deck and in weather deck.

INSPECTION

It is relatively easy to inspect pillars as they are accessible from the tween deck as well as from the lower hold.

It is important that the guides on which the wheels of the covers operate, or on which the covers slot-in, are free of any distortions as these can derail the cover on setting-down, or when they roll to close, whatever their design is.

The shafts around which the ends of the nonweathertight tween-deck covers rotate should be checked for deformations if difficulties in lifting and lowering of the covers are encountered.

LONGITUDINAL BULKHEADS
DESCRIPTION

Longitudinal bulkheads, to an extent, have already been discussed in Section "Pillars." There are no continuous longitudinal bulkheads of a solid construction in the cargo holds of dry cargo ships, either in the lower holds or in the tween decks, although these may be found in deep tanks.

A resemblance of solid longitudinal bulkheads exists if there are "shifting boards" fitted for the carriage of grain.

INSPECTION

The partial bulkheads erected at the ends of the tween decks are of limited height so visual inspection of welded fillets is relatively easy.

COFFERDAMS
DESCRIPTION

The structure of the cofferdam, in Figure 90 (photo on the LHS), extends between the tanktop of the ship (RHS photo, lower arrow) and the lower shelf plate, in Figure 93 (RHS photo, upper arrow).

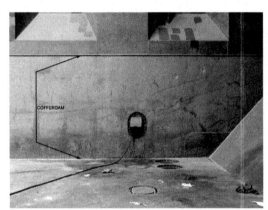

FIGURE 93

Cofferdam, LHS and its internal construction RHS.

Transversely, the cofferdam extends from port to starboard hopper tanks. It can be seen that the access manholes to this hold and the next have been removed.

INSPECTION

The internal spaces of a cofferdam can be accessed via manholes on each cargo hold (see Figure 93). Cofferdams may be used to store different liquids so they are subject to inspections and pressure testing.

SHELL PLATING
DESCRIPTION

The shell that makes up a ship's outer cover is what gives her the ability to float and so it is, perhaps, one of her most important parts. In addition, this shell contributes to the strength of the vessel. Generally, the thickness of the keel plate is substantial with adjacent strakes being of lesser thickness as their distance increases from the keel and from amidships.

To compensate strength at a distance from the ship's center the sheerstrake is of increased thickness and today its connection to the weather deck plating is by way of a rounder strake. Abrupt cuts and/

or openings are kept to a minimum as they introduce stress and strain concentrations that may lead to cracks in the plating. To further guard against these incidents any necessary openings are circular or elliptical.

INSPECTION

The external survey of the part of a ship's plating that is above her waterline can be visually inspected from the dockside, by utilizing a boat and from the quay side. The main purpose of this inspection is to record the presence of any indents and to identify their location, nature, and extent.

The internal inspection of the shell plating can be carried out visually, by accessing the tween decks and the lower holds. Bulk carriers and other ships that have double-skin construction will require an internal inspection so that the condition of stringers and other stiffeners can be ascertained.

TANKTOPS
DESCRIPTION

The tanktop and the outer-bottom shell form a very strong box-shape which is further supported by a lattice of longitudinal and transverse frames and floors. This type of construction gives the ship the capability of carrying not only bulk cargoes, but also large concentrated loads.

The loading/discharging equipment used by shore operators includes heavy grabs and small tractors, which can easily be accommodated on a large flat surface such as the one present in bulk carriers and general cargo vessels. This type of equipment, with the passing of time, will take its toll on the tanktops of a ship and invariably indents and set-down plating will occur (Figures 94–96).

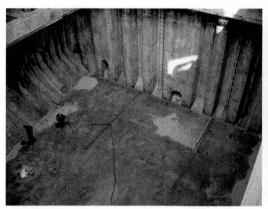

FIGURE 94

Bulk-carrier tanktops. The one on the LHS can accommodate containers.

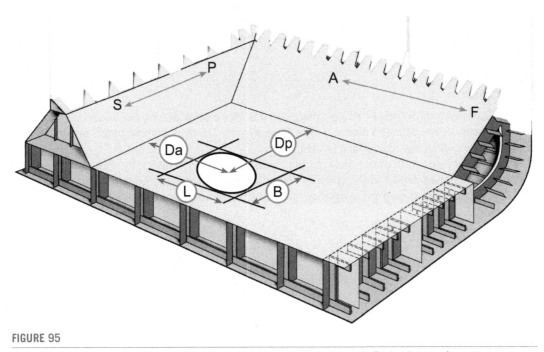

FIGURE 95

The plating damage (white ellipse) will be defined by its length 'L', its breadth 'B', its distance from the cofferdam/tanktop weld-seam 'Da' and its distance from the hopper-tank/tanktop seam 'Dp'.

FIGURE 96

Tanktop indentations.

Guidance:

Slight indents:	Upto 1.5 cm depth.
Moderate indents:	Over 1.5 cm and upto 2.5 cm depth.
Heavy indents:	Over 2.5 cm depth.

The water puddles at areas 'A' quickly reveal where the tanktop has dipped, or become depressed. This is because of the impact, or the weight of an external force (i.e. tractors or cargo grabs).

INSPECTION

Tanktops can be visually inspected by descending to them and taking careful notes of their condition. Tween decks ought to be first examined in their closed position and subsequently after opening so that all fittings can be observed, whether on the deck openings, or on the covers themselves.

CARGO HOLDS AND THEIR OUTFIT
DESCRIPTION
General cargo ships/tween deckers
Older tween Deckers and general cargo ships had steel cleats fitted in their frames, at tween-deck level, so that wood battens could be fitted and supported by them. This allowed cargo to benefit from improved air-circulation. In the same area there would be a number of ventilators providing either natural or forced ventilation of the cargo. A continuous centerline bulkhead, of heavy duty construction could also be found at tween-deck level.

The tanktops of the lower holds (depending on the ship's trading routes) could be fitted with container guides. In positions where the hull curvature made it impossible to stow a container at tanktop level, steel frames (forming stools) could be outfitted on which containers could be stowed, as a second tier level. Wood sheathing could be found covering the bulkheads and similarly heavy wood battens would be found fitted in the square of the hatch.

Refrigerated cargo carriers
This type of ship is constructed with a number of tween decks and a lower hold so that cargoes that require carriage at different temperatures and humidities can be simultaneously accommodated. Ventilation ducts are arranged on the port and starboard sides with the provision of hooks for the carriage of meat carcases (beef and lamb).

The actual deck surfaces are lined with aluminum lined profiles, which are holed at equidistant locations, so that refrigerated air can circulate freely and keep the cargo at the required temperature.

Ferries carrying cars and passengers
The lower holds of a number of ferries have been outfitted with longitudinal bulkheads so as to improve the stability characteristics of the ship in case of flooding. The bow incorporates a visor door that opens by lifting, whereas the stern incorporates a heavy duty ramp.

The decks of these ships are fitted with chain pockets where chains can be attached when cars need securing. Elevators or ramps are also used in this type of vessel to allow the movement and proper distribution of vehicles.

LO-LO, RO-RO, side-port vessels
Lift-on/Lift-off, Roll-on/Roll-off, Roll-on/Lift-off, and side-port vessels. The majority of these vessels have cargo holds which are almost 100% open (i.e., they have no concealed areas). The cargo holds have thus an advantageous shape (rectangular, open box) that allows the cargo to be stowed directly into the final stowage location.

Side-port vessels offer the advantage of utilizing conveyor belts, if the warehouse is located relatively near to the ship's side-door.

Bulk carriers

We have seen earlier that the cargo holds of bulk carriers provide for a rectangular shape cargo hold, where the sides are not completely vertical due to the presence of hopper and topside tanks, however, there are also double-skin bulk carriers which offer flush sides. Pad-eyes are found welded to the ships' frames so that wires and other means may be employed to restrain movement of different types of cargoes.

We have seen already that—should circumstances demand it—these ships can carry break-bulk cargoes, however, it is very important that the different blocks of cargo are properly separated (with colored, plastic netting) and that these cargoes are properly secured, so that they cannot fall off their stow positions when adjacently positioned cargo is removed for discharge.

INSPECTION

The surveyor will access the tween decks and the lower holds via ladders (mostly vertical type) with the exception of bulk carriers where the height of the cargo holds necessitates the employment of Australian regulation ladders with intermediate platforms.

Careful examination and detailed recording of any defects found is important, and all of these findings should be supplemented by clear photographs. A ruler, placed where a damaged part or area are located, will serve to give a quick comparison of how large the affected area is. Today, cameras are small and full of useful features (some even record sound) which the surveyor must take advantage of so that he can keep separate the various areas of the ship that he inspects.

SHIP PARTS: GROUP 3

HATCHWAYS AND HATCH COAMINGS
DESCRIPTION

Hatchways are a very important part of a ship's structure, and on a cargo ship they extend over at least 1/3 of the ship's beam. In container ships they are much wider and are only (approximately) 2 meters shorter in width as compared to the ship's beam (Figure 97).

The disadvantage of a ship's hatchway lies in the fact that—in effect—it is a very large discontinuity as far as the ship's deck is concerned. This reduces the strength of the deck so steps must be taken to reinforce the deck and overcome this negative effect.

This is achieved by arranging inserts (gusset plates) of greater thickness than the deck plating at the corners of the hatch opening. Figure 97 depicts the way these gusset plates (elliptical shape corners) extend in the hatch opening. These plates are of similar form in a tween-deck vessel. In bulk-carriers the protruding part of the gusset plate is covered by a shedder plate so that bulk cargo does not accumulate on these plates.

The hatch coaming extends in a fore and aft direction beyond the hatch opening, in the form of a bracket (see Figure 98 "A").

To provide additional strength, beams and longitudinal girders are arranged i.w.o. the hatch opening. In tween deckers, pillars are arranged at the corners of the hatchway opening to support the intersections of the beam and girders (Figure 99).

The height of the hatch coamings depends on whether they are located forward or aft on a ship's deck. At the fore location they are most likely to become exposed to green seas, accordingly they are of a greater height than the hatches in after locations, where they are better protected from adverse weather effects.

INSPECTION

Hatchways are the best inspected by approaching them from the weather deck and by carrying out a survey with the covers in their stowed position. By first inspecting the external condition of the stbd side of a hatchway, the next step would be to stand on the observation platforms (there are usually one at the after end and another at the forward end) and inspect the condition of the inner part of the hatchway on the port side.

This process can be repeated for each part of the hatchway. In the event that the stowed covers prevent the proper inspection of the forward and after transverse hatchways, these should be closed by the crew to expose the outer hatchways and their outfit i.w.o. the cross-deck structures.

The longitudinal connections of the hatchways to the topside tanks may be inspected by accessing the latter internally. Access the cross-deck to inspect the transverse connections of the hatchways at these locations. Always look for streaks of water or water stains on the inner surfaces of the coamings. This may be a sign that the hatch covers are not weathertight.

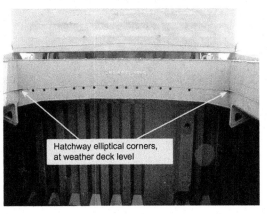

FIGURE 97

Openings of the hatchways.

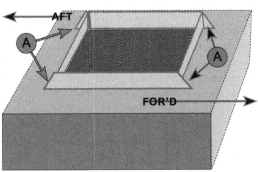

FIGURE 98

Hatchway elliptical corner-plates (gussets) at weather deck level. The diagram on the RHS shows the extended longitudinal hatch coamings.

FIGURE 99

Hatch-coaming extensions. The photo on the RHS shows hatch coaming and stays.

WEATHER DECKS
DESCRIPTION

We have already looked at the keel, double-bottom and side-shell plating, so by considering the ship's weather deck we complete the envelope that keeps her watertight (Figure 100).

FIGURE 100

Equipment that may be found on ships' decks.

In early designs the deck of a ship was given a slight curvature, called "camber"; however, modern ships have horizontal decks. All plating that forms a deck is supported by stiffeners, including web frames.

The plating i.w.o. any openings, such as hatchways, access hatches and in some instances around large diameter vertical piping is reinforced, this increased thickness is applied in order to counteract increased stresses, produced by the openings mentioned above.

Obviously, if it is anticipated that certain deck areas will be required to support large loads, especially point loads the plating is increased and the grade may also be changed.

Reinforcing of the weather deck may also be applied in areas where abrupt loads may occur, such as mooring ropes on bits (Figure 101).

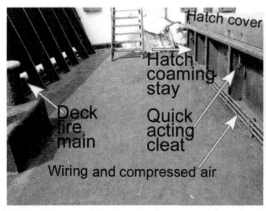

FIGURE 101

Equipment and piping found on ships' decks.

The description of the colors for piping onboard ships is given here for guidance purposes (Table 1).

Table 1 Pipe Colors	
Type	**Description**
Black	Waste media
Brown	Fuel
Gray	Nonflammable gases
Silver	Steam
Green	Sea water
Blue	Fresh water
Violet	Acids and alkalis
Red	Fire fighting
Orange	Oil (other than fuel)
Yellow	Flammable gases
White	Air in ventilation systems

In addition to the colors applied to pipes an arrow may be painted on the pipe to indicate the direction of the flow.

INSPECTION

Inspections of the weather decks are carried out by examination of the bulwarks (or rails, as the case may be), ventilators and sounding pipes and by noting the state of the longitudinal parts of the hatchways, including deck pipes, crossover steps and hatch or tank access manholes when going from forward to aft. It goes without saying that cross-decks and deckhouses will have to be made a part of this part of the inspection, as well as the forward and after transverse parts of the hatchways/hatch coamings.

The surveyor may also decide that this is the appropriate time to inspect the cargo-gear fitted on the weather deck. It is suggested that areas that are adjacent to the accommodation superstructures should be left to be examined separately.

BULWARKS
DESCRIPTION

Bulwarks are fitted for the protection of the crew and passengers (at the owners' choice bulwarks may be substituted by rails) (Figures 102 and 103).

1: Ventilator with Save all.
2: Mooring bitts.
3: Screwdown ventilator.
4: Roller chocks.
5: Screwdown ventilator.
6: Inspection platform with handle.
7: Welded attachment.
8: Welded pad-eye.
9: Lifting stanchion.

FIGURE 102

Bulwark at fo'c'sle level.

The horizontal arrows show the locations where the bulwark cap has been deformed and/or set-downwards.

At location '1' the bulwark has been interrupted to provide for a multi-angle fairlead.

FIGURE 103

Bulwark at weather deck level.

INSPECTION

The bulwarks (or the rails for that matter) do not add any strength to the ship's structure, but because they do get damaged during cargo loading and/or discharging operations they are definitely an item to be inspected. All indents and areas that are distorted/deformed must be recorded as they are claimed as "stevedore-damage" and they may require repairs.

MASTHOUSES AND DECK HOUSES
DESCRIPTION

In earlier ship designs, circa 1960-1970, were constructed with masthouses and deckhouses which were primarily used as a base for derrick and heavy derrick systems. Samson posts were usually arranged so as to pass through the plating of the masthouses and anchor themselves in the structure of the first deck they encountered (Figure 104).

FIGURE 104

Deckhouses serving as a base for heavy derricks.

Ventilators were often arranged to pass through the deckhouses/masthouses. If space permitted these constructions served to store lubricating oil and grease for the cranes and their equipment. Eventually, when derricks were replaced by cranes deckhouses became the supports of the crane-pedestals in addition to storage spaces.

INSPECTION

Because of their exposure to the elements their plating became subject to waste, with holes developing in places where the steel plating had become weakened. Railings at the tops of these deckhouses also suffered from similar effects. Other items of particular interest would be the condition of the deck stiffeners and the gaskets of the watertight deck doors.

A thorough survey would necessitate the accessing of the tops of these deckhouses via vertical ladders, welded on their sides.

ACCESS HATCHES AND ACCESS AND MANHOLES
DESCRIPTION

Access hatches can be found in many locations on the weather deck. The one on the left above is located on the fo'c'sle and provides access to the bossun's store, whereas the one on the right is located on the stern (Figure 105).

FIGURE 105

Examples of access hatches.

INSPECTION

There are two points every person using an access hatch must observe.

(1) *Before entering the hatch check and ensure that the cover of the hatch is locked in its open position.*
  - The locking bolt of the hatch on the right is missing its butterfly nut, so it cannot be held positively open.
  - Crew and other personnel accessing a cargo hold via a hatch customarily hold the top edge of the cover. Needless to say that if the cover is not held open securely, it will fall on the person within the hatch opening and can cause a serious accident/injury.

(2) *Before putting hands on the edge of the hatch those accessing the hatch must make sure that the same is not wasted, as this can cut someone's fingers.*

(3) *Ensure that the rungs of any ladders within the access hatch (and for that matter anywhere where they are being accessed for ascending or descending into a space) are in good order and condition.*

Check that the gasket on the cover is in good condition and that it remains effective.

So far, in this section, we have examined the construction and the points that concern the *Access Hatches* of a ship.

The second part of this section is concerned with the description and the inspection of *Access Manholes*. The difference between access hatches and access manholes is that the first are used quite frequently, especially when a ship is in port, whereas the latter are opened up only on particular occasions.

DESCRIPTION

Access manholes may be found located on a ship's weather deck, where they can provide access to the topside tanks and the upper stools of cofferdams. Manholes are also to be found in the cargo holds where they give access to the hopper tanks, the lower stools of the cofferdams and to the double-bottom tanks. They may also be found serving the fore-peak and after-peak tanks.

They are held in position by peripherally arranged bolts and nuts and they are made watertight by the use of rubber gaskets (Figure 106).

FIGURE 106

Manholes for accessing ships' tanks.

INSPECTION

Check the steel cover for deformation and then inspect the rubber gasket as to its reliability. Figure 106 on the LHS shows that one of the bolts is missing. This bolt should be replaced before the cover is placed in position and tightened. Similar remarks apply to the weather-deck manhole giving access to the topside tanks.

An important point here is to always check that the space you are about to descend into has been properly pre-ventilated and that it is fit for human entry.

HATCH COVERS (INCLUDING ULTRASONIC TESTING)

In older days ships were constructed with steel pontoon hatch covers which were made weathertight by using two tarpaulins, preferably three.

Wood wedges and steel bars were used to hold the tarpaulins in place. This type of cover could keep the water out of the cargo holds; however, the contribution of such covers to a ship's overall strength was very near to zero.

Steel hatch covers then hit the scene, introducing a concept brilliant in its simplicity. Their construction and general arrangement did contribute towards a better ship (when considering the aspects of a weathertight carrier, as well as a ship with improved strength, when compared to the old ships with pontoon covers), not forgetting the fact that they are easier and faster to open and close, requiring fewer crewmembers to accomplish this operation.

After the introduction of these new covers a period of trials begun and the lessons learned came taught naval architects and marine engineers alike that there were still problems to be solved.

Ships grew in deadweight size and in number of hatches while their freeboard was kept to the minimum height the international rules would allow.

Now let us consider the total assembly of the panels that make up one cover:

(A) *The panels are held down, peripherally, by the cleats.*

(B) *While the wedges take care of exerting pressure on the transverse joints.*

Also assuming that:

(C) *The forces produced by these two systems are uniformly distributed to all the joints.*

Also assuming that:

(D) *The maintenance of the coamings, of the covers and of all of their ancillary equipment is satisfactory.*

Then the whole combination of these parts makes up for an extremely rigid arrangement.

If we are to accept this statement as a given, then—with the vessel loaded and in *still water* conditions—the seal between the covers and the rest of the hull remains unbroken. So far so good.

However, once the ship proceeds to sea and encounters adverse weather conditions, her hull will be continually sagging and hogging which will be combined with a torque produced by other external forces.

It will be inevitable that the hull will flex and twist in such a way that the cleats and the wedges will not be able to hold the covers in a manner that would maintain the seal between the covers and the coamings.

There is an undisputed fact here and that is that the maintenance of the covers and the seals must be continuous and of the highest standard.

Nevertheless, today we may have to depart from existing ideas and theories and find a mechanism that allows the seal to remain independently fast on the coamings and on the covers, while it is afforded the opportunity to flex, deform and twist along its contact-path, without loss of its reliability as a *joint of two flexible media that remains uninterrupted* (Figures 107–110).

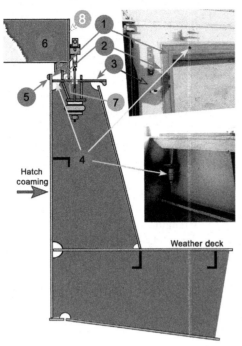

1: Hatch coaming plate.
2: Compression bar.
3: Flat bulb plate.
4: Drain hole, connected to drain valve.
5: Half-round bar, as coaming plate protector.
6: Patented rubber gasket.
7: Area where the cover side plate rests on the bulb plate '3'.
8: Cover side plate.

**Remark**: Balancing roller not shown.

Hatch coaming →

Weather deck

FIGURE 107

Diagram showing the seal between the hatch-coaming and the hatch-cover.

This photo shows a hydraulically operated hatch cover. (Hydraulic rams can be seen at the end of the hatch).
The underside of the cover is stiffened by longitudinally and transversely arranged web plates and beams, respectively.
The end of the cover carries deep beams, at its transverse ends.
Long rod cleats can be seen on the hatch-coaming.

A ladder on the longitudinal side of the coaming allows for inspection of the covers as well as inspection of the cargo when loading or discharging of cargo takes place.

FIGURE 108

Hydraulically operated hatch cover.

Source: MacGregor Finland Oy.

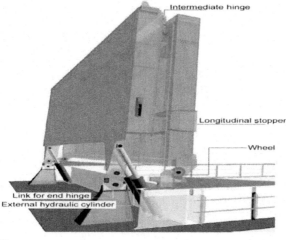

Intermediate hinge

Longitudinal stopper

Wheel

Link for end hinge
External hydraulic cylinder

This diagram shows what is, most likely, half a set of hatch-covers, in nearly stowed position at the end of the hatch coaming. The four hydraulic rams retract to allow the covers– by the action of the intermediate hinge - to roll to the RHS and take up their closed position. A similar action of the covers at the other end completes the closing of the entire hatch.

This is an efficient and fast way of opening or closing a ship's covers.

FIGURE 109

Diagrammatic of a hydraulically operated hatch cover.

Source: MacGregor Finland Oy.

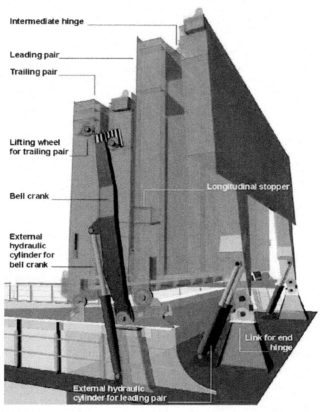

Intermediate hinge

Leading pair

Trailing pair

Lifting wheel
for trailing pair

Bell crank

External
hydraulic
cylinder for
bell crank

Longitudinal stopper

Link for end
hinge

External hydraulic
cylinder for leading pair

This diagram shows what is a full set of hatch-covers, in nearly stowed position at the end of the hatch coaming. The two bell cranks will allow the trailing covers to take a position at the far end of the hatch. In direct continuation the four hydraulic rams will retract to allow the leading pair covers – by the action of the intermediate hinge - to roll to the LHS and take up their closed position, next to the trailing pair. If this is a long hatch a similar action of the covers at the other end completes the closing of the entire hatch.

FIGURE 110

Diagrammatic of a hydraulically operated system, consisting of four hatch covers.

Source: MacGregor Finland Oy.

DESCRIPTION

Steel hatch covers made their first appearance in 1929 as a design by the MacGregor brothers. Since then, steel hatch covers have become the standard equipment for *General cargo ships*, *Bulk-carriers*, and *Refrigerated cargo carriers*. In addition to these hatch covers they design and build covers for other types of vessels.

Today, Cargotec MacGregor manufactures hatch covers to order, while at the same time they lease their designs to reputable third parties. There are also a number of other companies which manufacture hatch covers.

There are a number of different types of hatch covers and they fall into categories designed for those that are to be fitted on the weather decks of a ship and those to be fitted in a ship's tween decks.

The main types of hatch covers that are designed to be fitted on the *weather deck* of a deep-sea going, dry cargo ship are

(1) *folding type*
(2) *rolling type*
(3) *piggyback type*
(4) *reefer type*
(5) *stacking hatch covers*

The main types of hatch covers that are designed to be fitted on the *tween deck* of a deep-sea going, dry cargo ship are

(6) *folding type*
(7) *sliding type*

(1) *Weather deck covers—folding type.*

These covers are designed as high stowing types, with two, three, or four panels. These panels stow either at one end of the hatch or both. Single-pull compact-stow covers may have any number of panels. Their usage depends on the opening size of a hatch, the height of a hatch coaming and the length of stowage. The covers are operated by wire or by hydraulic means.

All covers are built specifically for one ship, depending on the requirements of the owners and the Class, as part of Load Line requirements.

Their configuration is for

(a) high-stowing/two-panel covers/twofold. These afford economical stowage, optimum hatch width.
(b) high-stowing/three-panel covers/direct pull. These are easy to install, with simple, economic operation of the ship's gear and they require minimum maintenance.
(c) high-stowing/three-panel covers/foldtite. These have the actuating equipment outside the panels of the hatch covers, with minimum stowage length and easy maintenance.
(d) high-stowing/four-panel covers/foldlink. These are suitable for long hatch coamings, with four or six panels stowing at each end. They feature minimum stowage length.

(2) *Weather deck covers—rolling type.*

These covers roll to stow on the sides of a hatch (side-rolling) and are mainly used by bulk-carriers. The usual arrangement is for one panel to stow on the port side, while the other stows on the stbd side. They roll on rails and they are driven by a rack and pinion mechanism. Alternatively, they can be driven by wires or chains.

The covers of this type offer simplicity of operation and independent panel-rolling. They can be made to be fully automatic. The external drive unit is always accessible with the panel closed and the operation is hydraulic, thus providing positive continuous and smooth movement.

(3) *Weather deck covers—piggyback type.*

Piggyback type covers can be fitted on a lot of different ship types. They stow one on top of the other and they can do so under fully automated conditions. They can stow on the ends or on the side of the hatch and can remain partially open. Their cleating can be fully automated.

(4) *Weather deck covers—reefer type.*

Reefer ships are designed to allow for the loading of 40 in. containers, with four containers being able to stow within the hatch opening.

The hatch covers are hydraulically operated, high-stowing folding type. These covers are easily operated and require little stowing-length.

Two panels stow on each end and their quick acting cleats can be manually or automatically operated. The sealing of these covers is important if the cargo holds are capable of maintaining a controlled atmosphere. The hatch covers have additional openings to facilitate pallet movements.

(5) *Weather deck covers—stacking hatch covers.*

Stacking hatch covers are of the multipanel type that can stow at the end of the hatch, but within the opening of the hatch. As an alternative the covers can be moved to any position on the hatch. These covers provide for fully automatic stacking at various locations. The number of covers that can be stacked in one location is, in theory, unlimited.

(6) *Tween deck covers—folding type.*

The hatch covers of this type fit flush so that the vessel benefits from a large area that can handle trailers, pallets and containers. The opening and closing of these covers is by means of hydraulic devices. Two panel covers are operated by wire arrangements, whereas four-panel covers are hydraulically operated. This enables the possibility of having the covers open on the port and stbd sides.

(7) *Tween deck covers—sliding type.*

The sliding hatch covers for tween decks can be fitted on both general cargo ships and in refrigerated cargo ships, especially where the decks are non-insulated. The panels slide, one under the other, as the hatch opens and they stow slightly sloping. Rack and pinion devices and wire arrangements can be employed in both the opening and the closing of these covers. These covers can be stiffened to carry containers, if this is required.

Having considered the various main types of hatch covers for weather decks and for tween decks, it is reasonable to say that the combination of (a) shipowners who ensure that the covers of their ships are frequently and properly maintained and (b) crews that have the knowledge and experience to operate hatch covers safely and in the appropriate manner, makes for a situation where cargo claims and indeed crew accidents can be kept to the minimum.

It is fair to say that ships do encounter adverse weather, especially in the winter months. Unfortunately, competing with nature is not a game that the hatch covers can win and damage to the cargo becomes an inevitability.

However, once there is a departure from these two prerequisites (a) and (b) above, we start to see how wrong things can go and how cargo claims can grow when ingress of sea water through the hatch covers happens. Damaged cargo claims, in 1965, averaged US $60K. In 2001, the average rose to US $243K.

Every year, insurers of ships and their cargoes pay substantial amounts of money towards claims of wet-cargo. It has been found that one of the main reasons of cargo suffering wet-damage is because of water entering the cargo spaces through the hatch covers of a ship. Accordingly, it is obvious that the keeping of hatch covers weathertight is one of the greatest concerns of shipowners, because if a survey proves that sea water entered the cargo holds through poorly maintained covers they will be liable to pay heavy fines, while their annual insurance premiums will increase.

Typical damage to hatch covers and hatch coamings is shown in Figures 111–113.

FIGURE 111

Gasket partially missing. Photo on the RHS shows a gasket out of place.

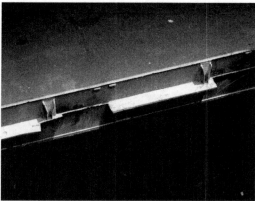

FIGURE 112

Gasket missing. Photo on the RHS shows the tween deck cover retainer plate damaged.

FIGURE 113

Tween deck cover retainer fractured. Photo on the RHS shows gasket missing intermittently.

INSPECTION

The damage to hatch covers may be related to

(1) *their rubber packing*
(2) *their steel structure*
(3) *their ancillary equipment*
(4) *rubber packing on the port and starboard longitudinal sides of the pontoons*
(5) *rubber packing on the for'd and after transverse sides of the pontoons*
(6) *condition of rising tracks and ramp stoppers*

With the covers partially open it is possible to inspect the condition of the retaining channels and the condition of the rubber gaskets. Once the covers are fully open the top transverses should be inspected. Not all covers are the same, so the surveyor, through experience and knowledge, will decide on the best mode of inspection in each particular case (Figures 114 and 115).

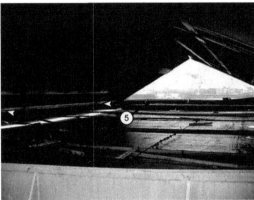

FIGURE 114

Suggested inspection mode.

FIGURE 115

Drain hole blocked and compression bar damaged. Some gaskets may be somewhat difficult to locate.

As "6" suggests (see Figure 114, p. 142) the inspection is not limited to the hatch covers. The coamings' flat will also need to be inspected for fractures and deformations, the compression bars will have to be checked for straightness (stand at one end—longitudinal or transverse—and look down the compression bar to its other end, to insure that it is absolutely dead straight) and the drain holes for not being obstructed/blocked.

Figure 115 (LHS), p. 142, shows the drain hole to be obstructed by debris and the compression bar to be wavy at its far end.

Figure 115 (RHS), p. 142, shows that there are gaskets and retention bars to be inspected when the covers are stowed upright.

If the surveyor's principals have requested that the covers should be tested in order to confirm that they are weathertight, there are a number of tests one could apply to this end.

(1) *Chalk test*

The compression bars of all the hatch panels are coated with chalk (i.e., longitudinal and transverse compression bars). The hatch covers are then secured, as though the ship was to be made ready for a sea-passage. The hatch covers, without any delay, are then opened.

The inspection involves the minute examination of the rubber packing. The chalk trace should be uninterrupted on all parts of the gaskets. If there are areas where the chalk-trace is not present, then these areas are marked as nonweathertight.

This type of test is not entirely conclusive and it is rarely used today.

(2) *Light test*

The hatch covers are secured, as though the ship was to be made ready for a sea-passage. The surveyor observes the periphery and the underside of the covers from inside the cargo hold, or the tween deck, with a view to pinpointing any location where daylight is visibly entering the dark space where the surveyor is standing.

This type of test is not entirely conclusive and it does need a sunny day for best results.

(3) *Hose test*

The hatch covers are secured, as though the ship was to be made ready for a sea-passage. The ship's fire main, on the deck, provides water to a hose. Hose testing is to be carried out with the pressure in the hose of at lease 2 bar (2 kgf/cm^2, 30 psi) during test. The nozzle is to have minimum inside diameter of 12 mm (0.5 in.) and located at a distance to the joint not exceeding 1.5 m (5 ft.). All of the seams and joints of the hatch covers are then subjected to a water hose test, while the surveyor is in the cargo hold or in the tween deck of the hatch where the test is being carried out.

Another surveyor, or assistant to the surveyor, ensures that the correct pressure is maintained and that the water is directed at the joints and seams of the panels, with the hose kept at about a meter away from them.

Once this exercise has been completed the covers are opened and all coaming flats and inner sides are minutely inspected for fresh entry of water. If there are areas where there is evidence of the water having gone past the gasket these areas are marked, as well as recorded to be nonweathertight. This process is repeated for all the hatch covers of the ship.

Care should be taken not to spill water onto the quay/pie. Some port authorities may fine a vessel if this happens. The water pressure mentioned earlier should be maintained during the test. A surveyor and an assistant are required. The test cannot be carried out with the vessel in the loaded condition as any leakage may cause wet damage to the cargo. It must also be mentioned that this is a test that takes considerable time to complete (Figures 116 and 117).

FIGURE 116

Inappropriate testing. The proper positioning is shown in the photo on the RHS.

FIGURE 117

Water stream directed to the flat plate. The photo on the RHS shows wrong positioning of the hose.

Figure 116 (LHS): Testing of a hatch cover as shown here is a complete waste of time and effort because the column of water, when applied at such increased distance cannot apply the required pressure.

Figure 116 (RHS): Applying the hose water as shown in this photo is the most appropriate way to test a cover.

Figure 117 (LHS): Here the water is applied to the offset bulb plate, hatch rest bar (not the seal path) and accordingly it cannot apply the required pressure.

Figure 117 (RHS): This is a variation of 117 (LHS) as the water, once again, is directed to the offset bulb plate, rather than the seal path. The crewmember ought to be standing on the deck and directing the water at the seal path.

Unfortunately, this method cannot establish with any accuracy where is the point where the water trace emanated from.

(4) *Ultrasonic test*

This test is based on the principle that ultrasonic sound produced within a cargo hold, can pass through any point where the compression bar and the gasket are not in full contact.

The hatch cover is battened down as though the ship was ready to sail. An electronic ultrasonic sound generator is then placed in the cargo hold and switched on (this means that the test can be carried out whether there is cargo in the hold or whether the hold is empty). A reading is taken with the equipment sensor at the open access hatch of this cargo hold which provides the so called open hatch value (OHV).

The person carrying out the test moves along the path where "compression bar and gasket" exists, with a probe capable of receiving the signal that passes through any opening created by an imperfect seal between the gasket and the compression bar. In this manner when the sensor receives a signal a leak is present and so its position can be established with accuracy.

Some instruments store a record of the locations where problems have been found to exist, whereas others make similar recordings in a manner that cannot be altered. The instrument used ought to be class approved.

Once the recording process has been completed the cover is visually inspected at the points where readings have given cause for concern.

Within Regulation 16.4 of the Load Line Convention 1966 states, inter alia, "…The arrangements shall ensure that the tightness can be maintained in any sea conditions…." However, testing almost always takes place with the vessel in drydock, or alongside a pier, or in a protected anchorage. In other words not under sea conditions a vessel encounters when crossing the North Atlantic in the winter months.

We have reached the end of this section. The following will serve to make up an "aide-memoire" of the main items to be inspected when holding an inspection of a ship's hatch covers.

Start at the after starboard side end of the hatch-coaming/hatch-cover arrangement—with the covers in their open position and securely stowed—and inspect the following:

- Extended bracket of the hatch coaming for signs of cracks or detachment from the deck.
- After end ramp and roller blocking devices.
- Drain valve under the hatch rest bar, at the after end of the hatch. Drain hole on the inside of the compression bar.
- Split joint auto cleat device.
- External hydraulic cylinder, or wire pull arrangements and sheave block(s), or crank and pinion mechanism.
- Split joint blocking device.
- Bell crank.
- Hatch-coaming plating and support stays, including any piping passing through openings in the stays, including longitudinal stiffeners.
- Chain condition, chain connections and chain carriers, if the chain is used.
- Eccentric wheels and balancing rollers.
- Spindles of quick acting cleats (they should not be bent) and rubber washer/ring at the lower end of the spindle (it should retain its elasticity).
- Junction pieces.
- Hatch inspection platforms.
- Towing arms and their wires.
- On reaching the for'd end of the coaming align your eye with the line of the compression bar and confirm that it is straight, free of any mechanical damage and clean. At the same time do the same for the coaming top and half-round bar.

- For'd end ramp and roller blocking devices.
- Drain valve under the hatch rest bar, at the for'd end of the hatch. Rain hole on the inside of the compression bar.
- Having taken all safety precautions access the top of the panels' channels and record their conditions (sometimes one can utilize an adjacent masthouse).
- Take remarks on the condition of all visible gaskets.

In a similar manner inspect the for'd transverse side of the hatch, the longitudinal port side of the hatch and the after transverse part of the hatch.

Once this part of the inspection is complete ask an officer to partially lower the after covers so that you can inspect their gaskets, compression bars and water draining channels. Record condition of the cover undersides. Carry out a similar inspection of the covers of the for'd end of the hatch.

In direct succession ask for the covers to be placed in their closed condition and check:

- cross joint wedge assemblies at the top of the cover-panels, including spring steel clips
- condition of top plating of the hatch cover-panels
- condition of any container fittings on the hatch cover-panels
- walk around the whole periphery of the cover and note the condition of the cover's side plating (Figure 118)

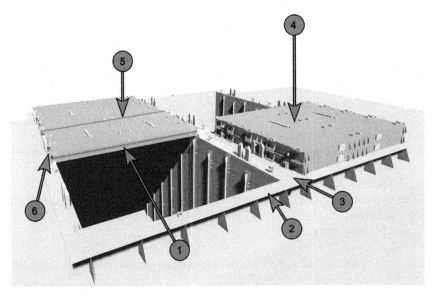

1: State of the compression bar.
2: State of the hatch rest bar.
3: State of the drain holes.
4: State of container fittings.
5: State of the compression bar and the water drainage system.
6: State of rollers and associated closing equipment.

FIGURE 118

Points to remember when inspecting hatch covers.

Source: MacGregor Finland Oy.

1. State of the compression bar.
2. State of the hatch rest bar.
3. State of the drain holes.
4. State of container fittings.
5. State of the compression bar and the water drainage system.
6. State of eccentric wheel assemblies and associated closing equipment.
7. State of the double drain channel, that is the channel between the compression bar and the protruding plating, also known as inboard rain channel, or peripheral drain channel.
8. State of the cover side plate.

Figure 119 shows that the vertical plating of the hatch cover not only supports the weight of the hatch cover but also controls the depth to which the compression bar presses the gasket of the hatch cover, which is housed within a channel bar. The wear of the hatch skirt [8] and that of the hatch rest bar [2] are important since they control the penetration of the compression bar [1] into the packing material. If such wear

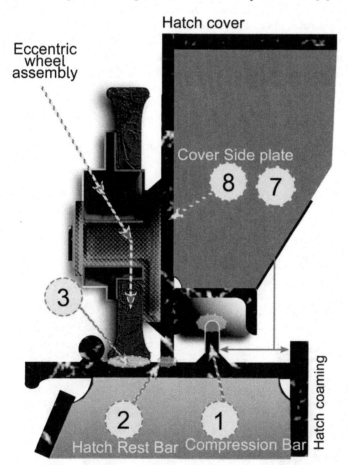

Hatch cover

Eccentric wheel assembly

Cover Side plate

8 7

3

2 1

Hatch Rest Bar   Compression Bar

Hatch coaming

Hatch covers are a complicated set of individual panels that are designed to work together and to remain in efficient service so long as there are no large deviations from the tolerances that they are designed for.

When a ship proceeds to sea, she will start running into heavy weather conditions which – one fashion or another – will upset the longitudinal and the transverse balance of the cover-panels to the hatch coamings. Even with new ships a combination of heaving, pitching and rolling will create situations where the seawater can enter the cargo spaces, as the ship's hull deforms in different ways as compared to her hatch covers.

FIGURE 119

Structural elements' wear affects the weather-tightness of the cover.

is allowed to progress unchecked, it will cause irreparable damage to the packing, but it will also impact on the overall balance of the cover, with further adverse consequences on the weathertight integrity of the pontoon-assembly as a whole. It is also true to say that a situation such as this will not only allow sea water to pass into the cargo space, but it will also weaken the overall strength of the rest plate. The flat hatch rest bar [2] is subject to wear and tear where the eccentric wheels roll, at [3].

Accordingly, the crew and owners must be vigilant to take care of all of these parts which need attention.

When a ship leaves her building yard all materials and equipment are new and free of any wear and tear. Usually, not much maintenance is required in the first few years of service, but then crew and owners must be vigilant to take care of the parts that need attention.

The points of contact between the hatch cover and the hatch rest should be placed at the top of the priority list of the candidates for both inspection and repairs (if this is found necessary).

As time passes the vertical cover plating lands repeatedly on the flat plate of the hatch coaming and this leads to the steel under the vertical plate and in areas at "1" and "2," to become worn, simultaneously there is wear on the vertical plate of the cover itself.

Soon the weather tightness of the seal will be lost and damage to the cargo will become only a matter of time (Figure 119). MacGregor Finland Oy has recently introduced the Electric Mac Rack rig. Appropriate explanatory notes have been provided in the captions of Figures 120, 121, 122, 123, 124, and 125 (pp. 148-151).

ELECTRIC MACRACK RIG

MacRack employs a combined rack and pinion drive and lifter system that makes separate hatch cover lifters obsolete.

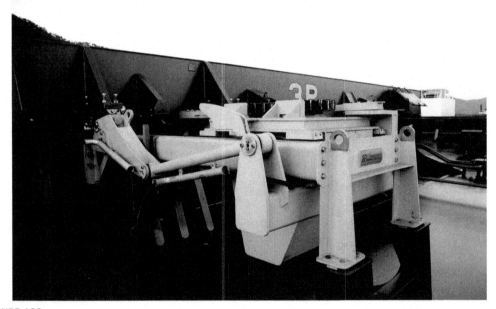

FIGURE 120

Electric MacRack rig (located in the middle of the port and stbd sides of the hatch coaming) starts lifting the cover.

Source: MacGregor Finland Oy.

FIGURE 121

Electric MacRack rig (located in the middle of the port and stbd sides of the hatch coaming). The cover wheel starts riding up a short rail. Cleats and stoppers automatically disengage.

Source: MacGregor Finland Oy.

FIGURE 122

Electric MacRack rig (located in the middle of the port and stbd sides of the hatch coaming). The lifting mechanism, in its up position, moves towards the cover. Rack and pinion engage. Transverse wheel is now at rail level.

Source: MacGregor Finland Oy.

FIGURE 123

Electric MacRack rig (located in the middle of the port and stbd sides of the hatch coaming). Cover about to stat rolling open.

Source: MacGregor Finland Oy.

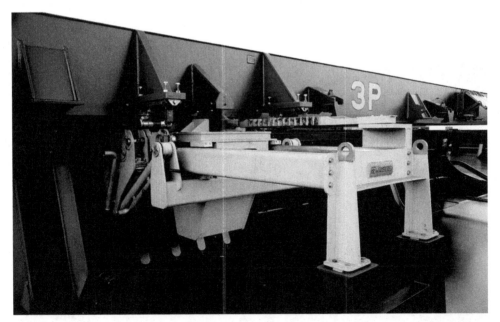

FIGURE 124

Electric MacRack rig (located in the middle of the port and stbd sides of the hatch coaming). The lifting mechanism, is in its up position, Rack and pinion starts rotating to open cover. Transverse wheel is now on the rail and rolling open.

Source: MacGregor Finland Oy.

FIGURE 125

Electric MacRack rig (located in the middle of the port and stbd sides of the hatch coaming). The lifting mechanism, is in its up position, Rack and pinion rotating to open cover. Transverse wheel is now on the rail and rolling open.

Source: MacGregor Finland Oy.

Electric operation removes the need for hydraulic pipework and other components. When opening hatch covers the lifting force needed is achieved by the MacRack lever mechanism, which converts rotational movement into vertical movement. When closing, the mechanism lowers the covers and pushes them together to achieve the correct compression of the hatch cover seals. This ensures the weathertightness which is vital for the protection of bulk cargoes (Figures 126–129).

BILGE ARRANGEMENTS AND TANKS
DESCRIPTION

Bilge water is a mixture of liquids that have leaked from the main-engine, generators, water separation processes from sludge tanks and drains from different cleaning systems. Bilge water collects in the bilge-wells of a ship, which are located in the lowermost part of the vessel just above the hull. Bilge water is created from flows that happen all the time and are, so to speak, predictable. However, it also emanates from leaks that take place intermittently and are not so easy to predict.

Today it is mandatory, in accordance with international legislation, that the bilge water that is discharged directly into the ocean from ships should not contain more than 15 ppm (parts per million) of residual oil in water. In this respect an oily water separator (OWS) is fitted on every ocean going ship

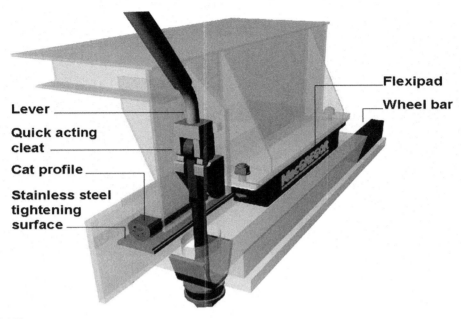

FIGURE 126

Seal arrangement. Quick acting cleat, cat profile, and flexipad arrangement.

Source: MacGregor Finland Oy.

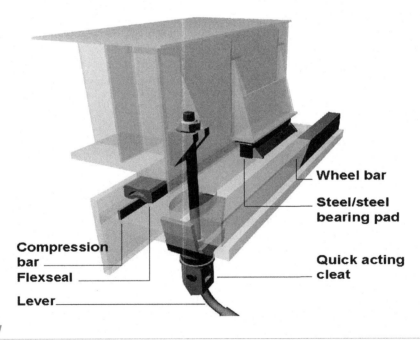

FIGURE 127

Seal arrangement. Quick acting cleat, flexseal, and steel to steel bearing pad.

Source: MacGregor Finland Oy.

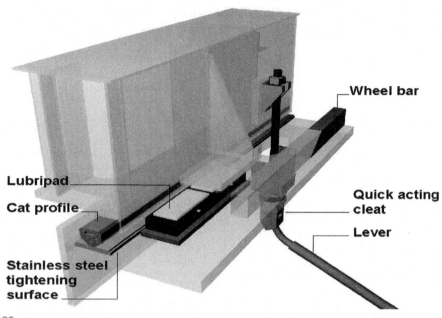

FIGURE 128

Seal arrangement. Quick acting cleat, lubripad, and cat profile.

Source: MacGregor Finland Oy.

FIGURE 129

Container deck stacking arrangement and container guides.

Source: MacGregor Finland Oy.

which takes care of this aspect. The bilges in the engine-room of a ship are located below the lowermost continuous platform. The latter accommodates the following pumps and tanks:

• Main sea water pumps	• Fuel oil drain tank
• Auxiliary sea water pumps	• Sludge drain tank
• Main ballast pump	• Lube oil drain tank
• Heavy oil transfer pump	• Lube oil renovating tank
• Diesel oil transfer pump	• Scavenger drain tank
• Lube oil transfer pump	• Stuffing oil drain tank
• Main engine lube oil pump	• Piston cooling water tank
• Bilge pump	• Bilge holding tank
• Sludge pump	
• Fire and general service pump	
• Fresh water generator ejector pumps	
• Piston cooling water pump	
• Jacket cooling water pump	

A bilge system on a dry cargo ship consists of a bilge-main in the ship's engine room which is connected to a pump, or pumps which draw from

- the sea
- cargo holds
- duct keel
- bilge main
- machinery spaces' bilges and overboard discharges

The pump(s) discharge overboard, to the fire main and to the OWS.

There are connections with the ballast pump and the general service pump so that three pumps can provide effective alternatives.

The bilge system (and for that matter the ballast system) is so constructed that it prevents the water passing from the sea or from water ballast tanks into the ship's cargo spaces, or machinery spaces, or from any watertight compartment of the ship to another.

Any connection of the bilge system suction to the sea, or ballast water system is fitted with a nonreturn valve or a cock that cannot be opened at the same time to the bilges and to the sea or the water ballast system.

Valves that can be locked or systems that can be fitted with blank flanges are used to prevent water from the sea entering deep tanks with cargo.

The amount of waste material that accumulates in a ship's cargo hold upon cleaning of the same as she comes to the end of a particular loaded voyage is surprising. Regrettably, part of this waste will eventually end up in the bilge wells where it might block the suction or the bilge sounding pipes, or cause erroneous results when these pipes are sounded. This is the reason for strum boxes to be fitted at the bilge suction end and these do keep the bilge suctions clear.

It is then left up to the crew to use whatever materials and tools are available to clean the bilge wells and remove the waste to the deck. The waste can then be removed to the pier for proper disposal (for additional comments—especially the *Important note*—please refer to p. 106 (Chapter 6).

It is important that any signs of blockage are investigated with a view to reinstating the bilge system in full working order. The oily waste together with other polluting substances like chemicals used for cleaning the machinery, is to be carefully removed in port so that it can be treated and disposed of in an environmentally safe way.

INSPECTION

In the cargo hold, remove the bilge well covers and inspect that space. It should be clean, dry, and free of any unwanted odors, especially if the ship is to load grain cargoes.

A bilge water separator is fitted with its own 15 ppm bilge alarm. However, when the local 15 ppm bilge alarm on a ship is broken beyond repair, the most appropriate action is to fit a new system. The reason OWSs are fitted with alarms and other devices, is so that the equipment will close when the capacity of the equipment has been reached.

SIDE SCUTTLES/PORTHOLES
DESCRIPTION

On a ship a side scuttle/porthole, when open, permits light and air to enter the quarters of the ship's crew that are located below deck. A side scuttle/porthole also provides a view to the sea. When closed, the side scuttle/porthole provides a watertight and weathertight barrier. The arrangement consists of a circular piece of strong glass contained within a metal circular frame. This frame bolts on the side of the ship's hull.

It is possible to have the glass of a porthole carried in a separate frame, hinged onto the base frame so that it can be opened and closed. In addition a side scuttle/porthole also has metal covers that can be securely fastened against the glass assembly, when necessary. The main purpose of the steel cover is to protect the side scuttle/porthole from heavy seas impacting on the hull.

INSPECTION

One of the telltale signs when inspecting a side scuttle/porthole are streaks of rust running vertically down on the hull plating. This is a definite sign that seawater leaks into the ship via the structure of a side scuttle/porthole (this being the reason this item has been included in the book). Corrosion may be so advanced that it poses serious concern for the safety of the crew. If this is the case, remedial steps should be taken immediately as side scuttles/portholes are usually located very near the ship's waterline.

SHIP'S OUTFIT: GROUP 4

DERRICK POSTS, CRANE PEDESTALS
DESCRIPTION

General cargo ships, bulk carriers, and refrigerated vessels share a common feature, that is the presence of derrick posts or crane pedestals (Figures 130–135).

Figure 130 shows a single post which serves to support the goosenecks of the for'd and aft derricks. Figure 132 shows a double post arrangement which serves to support the heel of a single derrick. In both instances, the motors that operate the wires as well as the controls of the derricks are located at the top of deckhouses. This is a good position for the derrick operator to be, since there he can have a good view of the hatch opening and of the hook progress while lifting or lowering cargo.

These posts are constructed strong enough to require no stays for support. The posts are usually of a circular section; however, square sections are sometimes also encountered. The arrangement of masts is carried to the weather deck and they are supported by a short bulkhead and a girder built along the centerline of the vessel and short transverse bulkheads on the port and stbd sides. Further stiffening is carried to the tween decks and sometimes even to the tanktop.

INSPECTION

In order to inspect derrick posts and crane pedestals, first carry out an external examination of the steel structure, making sure that this is sufficiently clean in order to allow a successful visual inspection. The same should apply to any deckhouses or masthouses on which posts and pedestals are erected.

Posts and pedestals should be free of any cracks as far as the tubular members are concerned. Make sure that all welding filets are equally free of discontinuities. With this opportunity, inspect access ladders from the weather deck to the top of masthouse and access ladders from the masthouse to the top of the main-post or the crane inspection platform.

With this opportunity, any motors and winches serving slewing and topping and cargo hoisting should be examined and witnessed to run properly. The control console should be included among all items examined.

Parts of the derrick post passing internally through the masthouse must be examined to prove that their connection to the ship's structure remain efficient and free of any damage. The same process applies to the foundation of posts and pedestals overall.

Equally, any service platforms at the top of the posts should be accessed so that the part of the post in that vicinity can be surveyed.

FIGURE 130

Post of two derricks (serving for'd and aft).

FIGURE 131

Cranes—two ready for work, one stowed on its stanchion.

FIGURE 132

Samson posts serving single heavy derricks.

FIGURE 133

Crane pedestal on top of deckhouse.

FIGURE 134

Crane pedestal on top of deckhouse. Boom stowing stanchion on RHS.

FIGURE 135

Two cranes on double base erected on a single-revolving pedestal.

DERRICKS, CRANES
DESCRIPTION

Single derricks have a capacity of 12 tons, whereas patent derricks—such as Hallen and Stülken—can lift up to 80-200 tons and 300 tons, respectively. While the lifting capacity of these systems is large, they have the disadvantage of being very slow.

The main parts and equipment of a ship's crane are as shown in Figure 136.

1. Cargo manipulation controls
2. Topping/slewing motor and winch
3. Cargo motor and winch
4. Topping/slewing motor and winch
5. Center post
6. "Y"-structure
7. Derrick boom
8. Gooseneck and gooseneck seat
9. Double hook and swivel

FIGURE 136

Hallen swinging derricks (post supports two derricks).

The SWL of the rig should be clearly painted on the side of the boom.

Today, derricks have given way to cranes because they are flexible and simple in their operation. In addition, a crane can access weights over a radius of 360°.

The common lifting capacity of ship-cranes is 25 tons. However, demands in the offshore industry have led to the design and construction of cranes with very large capacities in excess of 2000 tons.

Cranes are positioned on the ship's centerline; however, it is quite usual for cranes to be located on the port or starboard side of the deck, depending on the planned trade routes and cargo handling requirements of each individual vessel.

The main parts and equipment of a ship's crane are as shown in Figure 137.

The SWL of the crane should be clearly painted on the side of the boom.

INSPECTION

In accordance with class and government regulations, cargo lifting derricks and cranes are to be examined every 12 months.

Prior to commencement of the physical examination, the *register of lifting appliances and cargo handling gear* should be scrutinized, including all attached certificates. Here are two important points that the surveyor ought to make certain:

(1) There are no recurring problems.
(2) There are no outstanding recommendations from the previous surveys.

For derricks, the lists of blocks and reeving plans should be examined.

(A) The physical inspection of a crane should include the following equipment:
  * All loose gear
  * All ropes
  * Limit devices
  * Motors, winches, brakes, and drums
  * Sheave units
  * Hydraulic cylinders and pins (ram luffed cranes)
  * Jibs and heel pins
  * Slewing columns and machinery base
  * All bearings for slewing and bolts
(B) The physical inspection of a derrick should include the following equipment:
  * All loose gear
  * Ropes
  * Limit devices
  * Winches, brakes and drums
  * Deck fittings
  * Derrick boom
  * Mast fittings
  * Masts, derrick posts and guy posts

1: Slewing base, with slewing machinery internally.	7: Oil coolers.
2: Boom.	8: Electrical boards.
3: Pedestal.	9: Hoisting and luffing machinery.
4: Foundation.	10: Lifting wires.
5: Main Electric motor and pumps.	11: Lift block and swivel.
6: Luffing wires and rollers.	12: Double hook.
	13: Operator's cab.

FIGURE 137

Parts and equipment of a ship's crane.

Every 4 years the derricks or the cranes of a ship are to be tested as per the load shown below:

Safe Working Load (SWL)	Test Load
Up to 20 tons	$1.25 \times$ SWL
From 20 to 50 tons	SWL $+ 5$ tons
Above 50 tons	$1.1 \times$ SWL

Items subject to corrosion should be carefully examined. The surveyor may, at his discretion, ask for any excessive grease to be removed so that parts may be accurately and realistically inspected for any damage that might endanger life and property.

VENTILATION ARRANGEMENTS
DESCRIPTION

The purpose of "ventilation" onboard a ship is to provide a way of refreshing the air circulating in the accommodation, in the engine-room, and in the navigation bridge (Figures 138–139).

In addition, ventilation allows air to recirculate in the cargo holds in a way that there are sufficient air changes to maintain the cargo in good condition if the cargo requires to be ventilated or if it will benefit from it.

It is usual that cargo-hold ventilation is carried out by using a mechanically driven supply (e.g., electrical fan) and a natural exhaust system.

At the time cargo is loaded onboard, its temperature is the same as that which prevails at the point of loading.

However, during the voyage, the cargo will encounter temperatures that vary from that which existed at the time of loading.

Figure 140 LHS: *bad practice*. The ventilator cover packing has been completely covered in paint. This temperature difference will affect the cargo depending how warm or cold it is. If there is overheating the cargo will deteriorate, whereas cooling will cause the presence of condensation in the hold which will end up on the cargo surface. So it becomes clear why when passing to warmer climates it may be better not to carry out through ventilation with warm air.

Ventilators should be kept closed when moving from cold climates to equatorial ports and it should be arranged for the cargo to warm up very slowly (Figures 141–145).

FIGURE 138

ER ventilator suctions. The photo on the RHS shows a cargo-hold ventilator.

FIGURE 139

Different types of cargo-hold ventilators.

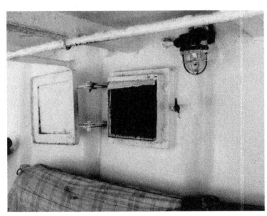

FIGURE 140

Accommodation fan. The photo on the RHS shows an accommodation ventilator.

FIGURE 141

Ventilators serving a ship's accommodation.

FIGURE 142

Ventilators serving a ship's accommodation.

FIGURE 143

Cargo-hold ventilators.

FIGURE 144

Cargo-hold ventilator.

FIGURE 145

Cargo-hold ventilators.

Having looked at the effects of temperature differences between the cargo and the air surrounding the vessel, it is important to mention that cargoes produce gases and odors which must be taken to the outside air quickly so that they may not be absorbed by some of the other cargoes in the same hold. Taint may cause some cargo to lose its value, this leading to cargo claims by the receivers (Figure 146).

Some cargoes not only require careful ventilation, but they will need the erection of air channels within the cargo stow in order to enhance the cargo ventilation.

It is a prerequisite that different cargoes are ventilated in different ways. So having a flexible ventilation system is only part of the solution since the ship's officers must have the required expertise to take advantage of it.

Essentially, the ventilation system provides fresh air to the cargo hold and it should be capable of extracting any unwanted smells and gases.

Equally important is the prevention of sweat, which forms on all surfaces in a cargo hold due to the inability of cool air to hold as much water vapor as warm air (Figure 147).

Sweat appears as *ship's sweat* and as *cargo sweat*. The former is a condensation that forms on the ship's sides and the latter is a condensation that forms directly on the cargo.

FIGURE 146

Cargo hold that carries containers and internal ventilation motors.

FIGURE 147

Cargo-hold ventilators located in the cross deck.

Ship's sweat will occur when the dewpoint temperature of a cargo hold is higher than the steel of the hold. Cargo sweat will occur when passing from cold to warm areas, so no inbound ventilation should be applied with extractors in operation.

Cargoes themselves may be hygroscopic (grain, flour, wood, cotton), which means that they are affected by atmospheric humidity. These cargoes cause ship's sweat, especially when passing from warm areas to colder areas. Sudden falls in temperature will cause ship's sweat (Figures 148–151).

FIGURE 148

DB tanks' ventilators.

FIGURE 149

DB tanks' ventilators.

FIGURE 150

Save-all for DB tanks' vents.

FIGURE 151

Save-all for DB tanks' vents.

Steel, machinery, and canned foods are nonhygroscopic cargoes. These cargos can be affected by cargo sweat. They rust and stain and also become discolored. If they have been loaded in cold climates and have proceeded to warmer climates, they will be affected by ship's sweat.

The purpose of ventilating a cargo is to avoid condensation and its adverse effects. Condensation comes into being when a cargo is loaded cold and then it is transported to a warm and humid climate. To better understand the mechanism of condensation we need to first look at what is "dewpoint" or "dewpoint temperature."

The air carries water as vapor. As the temperature of the air increases it will reach a stage at which it will no longer be able to hold its content of vapor water. The stage mentioned above is the Dewpoint temperature.

As a consequence of the foregoing, some of the water held as vapor will condense and will become liquid water. The amount of water vapor in the air at any given time is usually less than that required to saturate the air.

"Relative humidity" is the ratio of the partial pressure of water vapor to the equilibrium vapor pressure of water at the same temperature. Relative humidity depends on temperature and pressure of the particular environment.

A cargo is loaded and stored onboard a vessel in cold conditions. Subsequently, the vessel sails to an area where the weather is warm and humid. At this point in time, the cargo holds are ventilated with air of a higher dewpoint than the temperature of the cargo. The vapor held by the air will condense on the reels as drops of water.

This condensation will continue up to the time when the surface of the cargo becomes the same temperature as the dewpoint of the surrounding air. As an example see Figure 152, it shows that if the air is at 30 °C and the relative humidity is 80%, the dewpoint of this air is 26 °C. This means that no condensation will occur on a surface within this environment, that is at a temperature of 22 °C.

Water vapor in air is controlled by the temperature and the humidity. Air is said to be saturated when it has 100% humidity. Sea air has a humidity of 75%. When air is cooled it will eventually reach a temperature that makes it saturated.

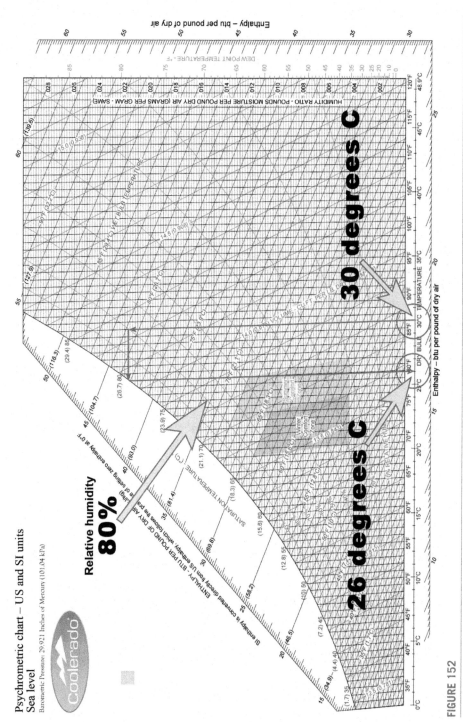

FIGURE 152

Typical psychrometric chart.

Source: Coolerado.

INSPECTION

DB tanks' air vent heads

Air ventilator-heads from ballast and FO (double-bottom) tanks are fitted with a mechanism that prevents the free entry of water. The same mechanism allows the passage of air or liquid to prevent excessive pressure or vacuum affecting the tank. They are fitted with air-louvers which can be removed for the inspection of the vent's internal parts. Some ventilator heads are also equipped with a wire mesh as a spark arresting screen on top of the tank.

It is a requirement to have the wire mesh for, e.g., heated fuel oil tanks or if there are anodes in a tank with a single air pipe or if there are specific national requirements.

Particular attention should be paid to the float and its rubber ring seals. If there is dislocation the unit's operation is impaired. The unit and its housing should also be inspected for corrosion. If a unit is to be replaced, the new part should be class-approved.

Accommodation and cargo-hold air vent heads

Fire dampers should operate easily and positions OPEN and CLOSED should be clearly marked. Internal access to the dampers should be provided as this is required for periodic inspection.

The fire dampers should be internally examined and proven structurally sound.

Dampers to be checked should be those for

(1) the cargo holds
(2) the machinery/pump room spaces
(3) the accommodation spaces
(4) the control stations
(5) the galley spaces
(6) other spaces

If there is evidence that the ventilation channels/trunking are blocked or leaking, it may become necessary to trace these and enter them with a view to finding the affected area and carrying out the necessary repairs.

Marine air conditioning

Heating, ventilating and air conditioning and refrigeration (HVAC&R) systems can be found on almost all ships from cargo vessels to luxury liners. While their function and operation share similarities with those in the built environment, their configuration is markedly different.

Marine air conditioning is a highly specialized area, where there is a need for innovative marine climate control solutions and support services to meet the needs of the shipyards, ship owners, and ship suppliers.

As the name implies, air-conditioning is defined as the treatment of air to modify the internal environment in terms of temperature, humidity, and fresh air content.

Marine air-conditioning systems reject heat from the space to be cooled into the surrounding seawater via the refrigerant flow in the sea-water cooled condenser. Vehicle systems operate in a very similar way to commercial systems—the heat is rejected to the outside environment via an air cooled condenser.

The air-conditioning process varies according to the application, from simple comfort-cooling to sophisticated total environmental control systems.

Some systems, such as chilled-water (CHW) systems on offshore accommodation vessels, feature extremely complicated distribution systems. These distribution systems measure internal accommodation pressures and manage space temperature via each diffuser via a computerized management system.

Through such a system, spill, return, and supply air volumes can all be managed. Ship HVAC systems are generally housed in designated compartments with the air handling unit (AHU), compressors, condensers, pumps, valves, and controls all together. And because ships can traverse the globe through varying climates—from the cold of the southern ocean to the tropics and everything in between—more is demanded of shipboard systems than would be of a fixed installation in a normal set-up on dry land.

As an alternative to air conditioning, the simplest system of ventilation operates exclusively with fresh air. The use of the air processing (heating or cooling) introduces economic and other factors which often justify the introduction of recirculation procedures.

VENTILATORS
DESCRIPTION

In Section "Ventilation Arrangements," we have seen numerous types of deck, accommodation, and double bottom tanks' ventilators.

Their descriptions have also been provided in the same section. This section shows the schematic of a 900 mm mushroom ventilator. The various internal parts that make up this equipment are shown in Figure 153. In the event that the height of the ventilator is greater than 900 mm the size of the deck stays is increased.

The hand wheel is turned clockwise or anticlockwise. This rotation, in turn, causes the weathertight cover to close or open. The wind cover may be attached to the air duct by means of bolts or by welding.

INSPECTION

Removal of the flame screen allows for the partial inspection of the ventilator. Other bolts allow the removal of the wind cover so that a detailed inspection and lubrication of the threaded rod may be carried out.

MASTS
DESCRIPTION

The construction of masts used to be based on tubular sections of steel. Today the designer's imagination is the limit, particularly with cruise ships where some really unusual designs can be found.

The masts have found their place on ships ever since there was a need to support the sails and lookout nests. In early designs, they used to be heavily stayed and they were made as composites of different parts.

The masts that we are describing here are not related to the single structures or the bipods that used to support derricks and other similar equipment. Figure 154 shows these masts carry equipment like radars and fog horns and navigation lights. Navigation lights are there in accordance with the rules preventing collisions at sea (COLREG).

Sometimes they even serve to support radio whip-antennas. In certain ships the masts are employed to support cargo-lights. Masts are constructed with a large base that tapers to a streamlined, smaller cross-sectional area, as its height increases. The entire mast structure is stiffened internally. Some

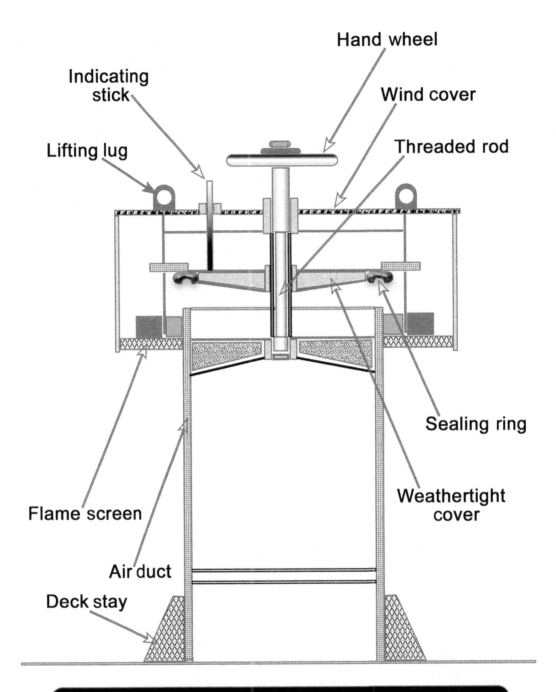

Hand wheel

Indicating stick

Wind cover

Lifting lug

Threaded rod

Sealing ring

Weathertight cover

Flame screen

Air duct

Deck stay

Structural diagram of a deck ventilator

FIGURE 153

Diagram showing the internal parts of a deck ventilator.

FIGURE 154

Ship masts.

FIGURE 155

Ships' masts.

modern ideas support the principle of providing sails on ships in order to reduce the consumption of hydrocarbons and to produce a greener concept (Figure 155).

If this is to be achieved there will be a requirement for masts to be incorporated in the ship's structure for the support of the proposed sails.

INSPECTION

The primary purpose of inspecting a mast should be to establish that there is no damage to its securing on a deck or other ship structure. In the event that the mast is stayed, the wires used for this purpose ought to be examined to establish that they are in good condition and that they are adequately greased.

It is also important that all the appropriate lights are on the mast and that they are in good order. Attention should be paid to all parts of the lantern including the provision of electrical connections, the lamp-holder and the lens.

Navigation Light Controllers, which provide a means of control and monitoring of the status of the navigation lights onboard the vessel to the Officer of the Watch, should be tested.

CONTAINERS

DRY CONTAINERS
DESCRIPTION

Moving cargo from one port to another, across the world's seas, can be challenging. Transport documents, regulations, country-specific compliance, local requirements, and customs clearance are just some of the steps to be considered before your cargo will arrive at its final destination. Proper export documentation and import licenses and requirements are becoming increasingly important factors when transporting cargo. Dry containers are constructed in lengths of 20′ and 40′ (general purpose), and they are manufactured from either aluminum or steel. They are suitable for most types of cargo. Aluminum dry containers have a slightly larger payload than steel, and steel dry containers have a slightly larger internal cube.

There are numerous designs of shipping containers, but 95% are what are termed "dry van containers"; these carry general freight, and they have the following general characteristics:

- They ought to be built to a full marine specification—manufactured from corrosion-resistant Corten steel.
- They ought to be protected with marine grade paint.
- They ought to be fitted with additional ventilation—to ensure condensation is not a problem.
- They ought to be fitted with high locking bars for ease of use.
- They ought to have a 28-mm marine ply floor that is specially sealed and certified as being rot and vermin proof.
- They ought to have a factory fitted high security lock box which shrouds the padlock and protects it from interference.
- They ought to be capable of being stacked.

Dry containers come in several sizes:

20′ with payloads up to 28.2 metric tons

40′8′6″ with payload up to 28.8 metric tons

45′9′6″ high with a capacity of 85 cubic meters

Pallet Wide Containers: This is a container type which is used for standard Euro Pallets (1.2 m × 0.8 m). These containers have 2.45 m internal width, which allows companies to load 5 additional pallets (10 if they can double stack).

Ventilated Containers: From the outside, these containers seem identical to dry containers. They have an internal ventilation system which accelerates and increases the natural convection of the air within the container. It's mainly used for organic products that require ventilation. These can be used for anything from coffee beans in bags to copper wires. Flexi tank is an additional tool that can be used with a dry container for carriage of different products.

Open Top Containers: These are just like dry containers; however, their top is open. Either they have a removable top or they are covered by a tarpaulin (especially, for outgauge cargo). They are mainly used for cargo that is too long or too heavy to be loaded by a forklift from the door. These units are also for cargo that is of a height exceeding the height of the container. Besides being used for machinery and large finished goods, it is also used for block marbles, glass, and bulk minerals.

Flat Rack Containers: Flat rack containers are mainly used for over height, over width, or heavy cargo. They have collapsible sides which help to load even longer pieces. Some of the flat racks can carry as much as 45 tons per container. From machinery to yachts, you will see different cargos moved on this equipment.

Platform Containers: These are pretty similar to flat rack containers. Used for heavy and/or overwidth/height cargoes. An interesting way of using a platform containers is, when laid side by side on the deck or in the hold of container ships, they can be used to transport even larger noncontainerizable cargo.

Tank Containers: These containers are mostly used by tank operator companies who are renting their equipment globally. The equipment has a 20′ frame with a tank inside. They are used for both nonhazardous and hazardous liquids.

INSPECTION

Dry containers usually suffer mechanical damage either on loading onto or discharging from the overseas carrier. They may also be damaged during manipulations in the container yard. If the damage is restricted to the sides or the roof, the damaged portion is cropped off and renewed.

If the damage has affected the corner posts or other important parts of a container, structural integrity repairs will be carried out under the supervision of an accredited repairer and the unit may need to be weight tested on completion of the repairs.

An external examination will involve the condition of the roof of the unit, which can be accessed by a small height ladder. In direct continuation, enter the container and have the doors fully closed (make sure that an assistant remains outside so that he can let you out once your internal inspection is complete). In the dark interior look for any light that might be entering the cargo space from the outside. If you do find such points of light entry, mark them with chalk and exit the unit (Figures 156–158).

Forty foot dry container.

FIGURE 156

Dry box (40′).

Internal view of a dry container.

FIGURE 157

Floor of a dry container in the process of being cleaned before being loaded with its next cargo.

Container locking arrangement with a numbered seal.

FIGURE 158

Despite their locking arrangement, containers have been targeted by gangs interested in their contents.

REEFER CONTAINERS

DESCRIPTION

A reefer container when inspected from its rear, where the doors are located, looks the same as a dry container. It is when we walk to the front of the unit (the end opposite to the doors) that we are presented with a combination of equipment and machinery.

Closer examination shows the following to be present: (1) reefer compressor, near the base of the container; (2) air exchange controls on one side; and (3) digital circuit box on the other side. This regulates the temperature at which the cargo is kept, including the humidity.

In the same area, there is a plug which allows the download of the container electronic data to a mobile unit that can then be downloaded the data to a PC for appraisal and storage.

The cargo space floor is of a T-profile. The air, once refrigerated by the container's machinery, rises through the perforated floor and cools the cargo. This is known as bottom air delivery.

Certain cargoes, particularly vegetables and fruits, produce gases (as they continue to breathe and produce heat in the container) which—if not controlled—can lead to damage to the cargo.

These gases can be removed by the ventilation and the circulation of fresh air in the cargo space. Once the air has been circulated through the cargo, it returns to the machinery space of the container where it passes over the evaporator coil and is cooled in this way.

Reefer containers operate in two modes:

1. Frozen mode temperature control is accurately achieved through the return air.

   In the case of frozen cargo, air has to flow around the cargo so there should be no gaps between the cargo and the walls and the cargo itself, so the cargo has to be block stowed.

   Frozen cargo is regarded as "inert" and is normally carried at or below $-18\,°C$ ($-0.4\,°F$).
2. Chilled mode temperature control is accurately achieved through the supply air flow.

   In the case of chilled cargo, air has to flow through the cargo at all times so that heat and gases are removed; therefore, the cartons used should have ventilation.

   Many chilled cargoes (e.g., fruits) are regarded as a "live" cargo since they continue to respire postharvest and as such are susceptible to desiccation (wilting and shriveling).

   This is not the case with commodities such as chilled meat or cheese. The minimum fruit carriage temperature is usually not lower than $-1.1\,°C$ ($30\,°F$).

   It is important to note that a reefer unit is not designed to reduce the temperature of the cargo but rather to maintain the temperature to which the cargo has been precooled.

INSPECTION

The inspection of a refrigerated container must follow the requirements as well as the recommendations of the individual entities that have contributed to the construction of such units. This means that:

(A) The manufacturers of the box itself ought to produce a set of rules and recommendations relevant to the proper upkeep of their equipment.
(B) The manufacturers of the refrigerating machinery and of the parts immediately related to it (i.e. condensers, evaporators, fans, etc.) ought to produce a set of rules and recommendations relevant to the proper upkeep and setting up of their equipment.
(C) National and international bodies, responsible for the carriage of foods, fruits, etc., in refrigerated, in cooperation with the growers, and the shippers of such commodities ought to produce their own appropriate rules and regulations.

Once the parties included in (A), (B), and (C) above have produced their guidelines and rules, the next natural step would be for the owners of the refrigerated containers to prepare a document that would follow these guidelines and rules on every occasion a reefer unit is about to be loaded with a new cargo consignment. A pretrip inspection (PTI) is very much the result of this synergy, which is followed by the various shipping companies and by their reefer container land-based counterparts.

REFRIGERATED (INSULATED) CONTAINER INSPECTION LIST

External Inspection can be carried out by a reefer yard superintendent, a ship officer and/or a marine surveyor, who can record information, or carry out tests and inspections of the following components, whilst recording their condition.

1. **Corner Fittings:** There are eight (8) corner fittings located at the upper and lower corners of the container. These fittings provide a means of stacking, manipulating, and onboard-securing.
2. **Top-End Transverse Member:** There are two (2) transverse structural members, at the top of the front and the rear of the unit. They connect the corner fittings. The top-end transverse members above doors are known as "door headers".
3. **Bottom-End Transverse Member:** There are two (2) transverse structural members at the bottom of the front and the rear of the container joining the bottom corner fittings of the end in question. Where mounted below end doors, these members are commonly known as "door sills".
4. **Top Side Rail:** There are two (2) longitudinals at the top side of a container, on the right and left sides. These join the top corner fittings of the side, at the front and at the rear.
5. **Bottom side Rail:** There are two (2) longitudinals at the bottom side of a container, on the right and left sides. These join the bottom corner fittings of the side, at the front and at the rear.
6. **Corner Post:** There are four (4) vertical structural members. They are located at either side of the front and on the rear of a container. They join the top and bottom corner fittings.
7. **Floor:** The floor is the part of the container which supports the weight of the cargo within it. The floor is constructed from aluminium panels incorporating floor components which are specially designed so as to allow air (or gas) to pass under the cargo.
8. **Floor Bearers:** These are transversely arranged components, within the "base structure" of a container, which support the floor. In certain cases, there may also be longitudinally arranged strength members.
9. **Roof Bows:** These are transversely arranged members under the roof plating, forming part of a rigid roof structure.
10. **Fork Lift Pockets:** These are reinforced openings/pockets, running transversely across the "base structure" of the container. They penetrate the bottom side rails at predefined locations, and they so permit the entry of the tines of forklifts so that the unit can be lifted and manipulated as necessary.
11. **Gooseneck Tunnel:** A recess (at the front) in the bottom of a container having specified features that permit the use of a single tine/gooseneck-arm to enter and lift the container.
12. **End Doors:** These are panels located on the rear wall of a container. They are load-bearing, and they open or close so as to allow the loading and the securing of cargo in a container.
13. **Roof:** This is a rigid plating, usually corrugated and weatherproof, which forms the top of a container. It is bounded and supported by the top-end transverse members and the top side rails.
14. The triangular area within the RHS door panel contains the following information: Owners' code, serial number, check digit, country code, size/type code, maximum gross weight, tare weight, maximum useful load.
15. The rectangular area within the bottom of the RHS door contains the following information: CSC plate, Owners' plate, Customs' approval plate.

16. The area at the forward upper end of the RHS side-plating carries various approval plates.

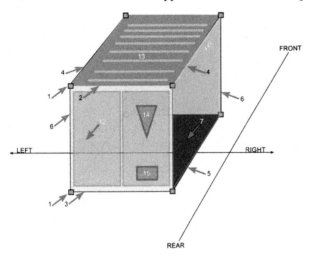

The inspection of refrigerated containers may also include the following items: cargo chamber insulation and plating, drains, drain protection plates, gratings, and air-baffles (and their supports); Doors' insulation, hinges, locking bars, handles, seals, and rubber stoppers; Cargo load line marks; Cargo hooks. On the front settings and digital readout, or Partlow chart should be examined. The position of the air intake flaps ought to be noted.

The attending inspector ought to record the machinery type and manufacture date whilst recording the refrigerant type. If the unit is fitted with battery packs, a suitable note ought to be made. The unit cargo-chamber is to be inspected for cleanliness, odors, dirt, and the presence of free running water. The unit's doors and gaskets, baffle plates must be found to be present and tight. Any internal and/or external impact damage is to be recorded. Inspect the unit and its machinery per accepted procedures, but without opening of parts ordinarily concealed. Make a note of the ambient temperature at the time of the inspection.

Refrigerated containers have cooling facilities built into them. The internal air circulation is regulated so as to ensure the maintenance of appropriate temperatures within the cargo compartment. Cold air is constantly circulated through and around the cargo space to dissipate any heat present. Cold air is delivered into the cargo compartment at the bottom, through "T" extruded floor-plates/gratings. Once this cold air has served its purpose, it is drawn off again below the container ceiling. The circulating fans then force the air through the air cooler, which also acts as the evaporator in the cold circuit, and back through the gratings into the cargo. The front of a refrigerated container contains all of the machinery that cools the cargo, including the means of regulating the admission of fresh air and the exhaust of air that has circulated in the cargo space (See Figure 159.) The digital controls of a refrigerated container are also located at its front and these have the means to download the performance data of the unit over long periods of time. (See Figure 160). The air valves, regulating the input of fresh air as well as the removal of air that has cooled the cargo are also located at the front of the container (See Figure 161). A tank container, built to carry gases under pressurized conditions, is shown in Figure 161. Different manufacturers of such containers select the location of their valves and other related equipment serving the pumping of gas in and its discharge.

Readers are recommended to visit: Source: HAMBURG SÜD

http://www.hamburgsud-line.com/hsdg/en/hsdg/reefer_7/equipment_1/equipment.jsp

This site contains a wealth of data, covering both refrigerated containers and reefer cargoes (Figures 159-162).

FIGURE 159

Front of refrigerated container.

FIGURE 160

Control panel of a refrigerated container.

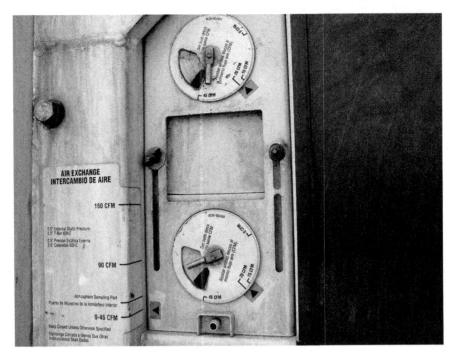

FIGURE 161

Refrigerated container air exchange controls.

FIGURE 162

Tank container.

CONTAINER SECURING AND OTHER ASSOCIATED EQUIPMENTS
DESCRIPTION

Here, we shall look at container stowage and lashing systems. These systems are subject to continual improvements which are dictated by the ongoing demands of shipowners for faster loading and discharging vessels. These systems are also subject to the stringent safety regulations for the stevedores. It goes without saying that all of this equipment is subject to Classification Societies' approval.

Below follows a list of the various pieces of equipment that are used for stacking and securing containers onboard ships:

Source: MacGregor Finland Oy.

- Lashing bar

- Lockable stacking cone

- Twinlocks and twistlocks

- Twinlocks and twistlocks

- Twinlocks and twistlocks

- Twinlocks and twistlocks

- Turnbuckles

- Pressure fitting

Other equipment used for the stacking and securing of containers is as follows:

- Interbridge stacking cones
- Pressure adapters and antirack spacers. Lockable stacking cones
- Bottom stacking cones
- Twinlock and twistlocks
- Bottom twistlocks
- Pressure adapters
- Linkage plates
- Twistlock operating rods
- Bridge fittings
- Turnbuckles
- Lashing bars
- Hooks (Figure 163)

FIGURE 163

Container lashing bars, D-rings, and raised sockets.

Source: TEC container.

INSPECTION

The best way of dealing with loose fittings like those described above is to create a list of descriptions and a small diagram or photo next to each piece of equipment so that they become readily identifiable. It is then easy to go about the ship's holds and decks and record the numbers of each different type of fittings.

These fittings can be found stored in bins between the stays of hatch coamings and in recesses in the lower holds. These fittings are subject to impact damage and corrosion effects, so they should be inspected when placed in bins and should be taken out of use if there is doubt as to their being fit for purpose.

SURVEY OF BUNKERS R.O.B.

INTRODUCTION

BUNKER FUEL

In the days when steam-powered ships ruled the waves, coal-fired boilers produced the steam that turned their propellers. This fuel was stored in what was then known as coal bunkers. The name "bunkers" has survived until today, but now bunker fuel tanks are used for the storage of petroleum products which are today's favorite fuel.

Broadly speaking, a bunker fuel is defined as the fuel that powers the internal combustion engine of a ship and it is bunker C or No. 6 fuel oil. This is the most common "bunker fuel" and it is often a synonym for No. 6 fuel oil.

No. 6 fuel oil is also commonly called "Heavy Fuel Oil" (HFO), and it is a high-viscosity residual oil requiring preheating to 104-127 °C. The residual is the material remaining after the more valuable cuts of crude oil have boiled off. The residue may contain various undesirable impurities including 2 percent water and one-half percent mineral soil.

The high viscosity requires heating, by a steam system, before the oil can be pumped from a bunker tank. Accordingly, residual fuel oil is less useful because it is so viscous that it has to be heated with a special heating system before use and it contains relatively high amounts of pollutants, particularly sulfur, which forms sulfur dioxide upon combustion. Power plants and large ships are, however, able to use residual fuel oil.

In recent years, marine engines have been designed to use different viscosities of fuel. The unit of viscosity used is the Centistoke.

Fuels most frequently used are:

Intermediate fuel oil with a maximum viscosity of 380 Centistokes (<3.5% sulfur)
Intermediate fuel oil with a maximum viscosity of 180 Centistokes (<3.5% sulfur)
Low-sulfur (<1.0%) intermediate fuel oil with a maximum viscosity of 380 Centistokes
Low-sulfur (<1.0%) intermediate fuel oil with a maximum viscosity of 180 Centistokes
Marine diesel oil
Marine gasoil
Low-sulfur (<0.1%) Marine Gas Oil—the fuel is to be used in EU community Ports and Anchorages

Ultra Low Sulfur Marine Gas Oil—referred to as Ultra Low Sulfur Diesel (sulfur 0.0015% max) in the US and Auto Gas Oil (sulfur 0.001% max) in the EU. Maximum sulfur allowable in US territories and territorial waters (inland, marine, and automotive) and in the EU for inland use.

Density is another parameter quoted with regard to fuel oils since they are purified before use in order to remove water and dirt. Centrifugal force is used by purifiers onboard, and the oil must have a density which is different from water.

	Density at 15 °C (kg/m³)—Max	Viscosity at 40 °C (mm²/s)—Max
Marine distillate fuels		
DMX	–	5.5
DMA	890.0	6.0
DMB	900.0	11.0
DMC	920.0	14.0
Marine residual fuels		
RMA 30	960.0	30.0
RMB 30	975.0	30.0
RMD 80	980.0	80.0
RME 180	991.0	180.0
RMF 180	991.0	180.0
RMG 380	991.0	380.0
RMH 380	991.0	380.0
RMK 380	1010.0	380.0
RMH 700	991.0	700.0
RMK 700	1010.0	700.0

ASSESSMENT OF THE WEIGHT OF PETROLEUM PRODUCTS
MASS

Mass (m) is a measure of the amount of matter that occupies a body, and it can be considered in terms of the gravitational field produced by the body; in these circumstances, it is referred to as *gravitational mass*. Mass is also a quantitative measure of the inertia of a body (inertia is the property causing resistance to change the motion of a body); accordingly, sometimes it is referred to as *inertial mass*.

The *force (F)* on a body and the resulting *acceleration (a)* are expressed as:

$$F = m * a.$$

Mass remains constant, unlike the volume and weight of a body; however, Einstein's special theory of relativity indicates that the mass of a body does not remain constant but increases with *speed v*:

$$m = \frac{m_0}{\sqrt[2]{(1 - v^2 / c^2)}}$$

where:

m_0 is the mass at rest
c is the speed of light
Unit of Mass: [SI] kilogram (kg), [Metric, technical] kilogram (kg)

WEIGHT

Weight (W) is the force exerted on matter by the gravity of the earth. The weight of an object of *mass (m)* is equal to:

$$W = m * g$$

where:

g is the acceleration of free fall ($9.80665\,m/s^2$)
Unit of Weight: [SI] Newton, [SI] kg*m/s^2

An object of 1 kg mass, lying on the surface of the earth, weighs 9.80665 N.

VOLUMETRIC MASS DENSITY

Density (ρ) is the mass per unit volume of a material, and it varies with temperature and pressure (therefore, these should be stated).

$$\rho = \frac{m}{v}$$

Unit of Density: [SI] grams per cubic centimeter [Metric, technical] kilograms per cubic meter

In some cases, in the oil and gas industry, density is loosely defined as weight per unit volume (although this is scientifically inaccurate). This density is more appropriately called *specific weight*.

Density is sometimes replaced by *relative density* or *specific gravity*. These are dimensionless quantities, and they represent the ratio of the density of a material to the density of water, at 4 °C (or some other specified temperature). The reciprocal of the density is occasionally called *specific volume*.

OTHER DEFINITIONS

Clinkage: Oil residues which adhere to the surface of a tank wall and structures on completion of discharge.

Volume correction factor: A factor dependent upon oil density and temperature which corrects volumes to the standard reference temperature. Such factors shall be obtained from the latest API-ASTM-IP Petroleum Measurement Tables.

Total observed volume (TOV): The volume of oil including total water and total sediment, measured at the oil temperature and pressure prevailing.

Gross observed volume (GOV): The volume of oil including dissolved water, suspended water, and suspended sediment, but excluding free water and bottom sediment, measured at the oil temperature and pressure prevailing.

Gross standard volume (GSV): The volume of oil including dissolved water, suspended water, and suspended sediment, but excluding free water and bottom sediment, calculated at standard conditions, for example, 15 °C and 1.01325 bar.

Total calculated volume (TCV): The gross standard volume plus the free water measured at the temperature and pressure prevailing.

Wedge formula: An equation relating the volume of a liquid material in a ship's tank to the dip, ship's trim, dipping point location, and the tank's dimensions when the ship's calibration tables cannot be applied (please refer to example in Appendix F).

Weight conversion factor: A factor dependent on the density for converting volume to weight-in-air. Such factors shall be obtained from the latest API-ASTM-IP Petroleum Measurement Tables.

Weight (products): Weight-in-air: The weight of oil, excluding free water.

The International Maritime Organization (IMO) has recently decided that ships need to consume fuel, which is cleaner than HFO and not as harmful. Ships trading in designated emission control areas will have to use onboard fuel oil with a sulfur content of no more than 0.10% from 1 January 2015, against the limit of 1.00% in effect up until 31 December 2014.

The emission control areas established under MARPOL Annex VI for SOx are: the Baltic Sea area, the North Sea area, the North American area (covering designated coastal areas off the United States and Canada), and the United States Caribbean Sea area (around Puerto Rico and the United States Virgin Islands).

BUNKER SURVEY EXECUTION AND PRACTICAL CONSIDERATIONS
CASE 1: FUEL SUPPLIED TO AN OCEAN TRADING VESSEL BY A BUNKER BARGE

A surveyor attends onboard when the ship is about to receive either FUEL OIL and/or DIESEL OIL. The purpose of this attendance is to sound the tanks of the bunker barge as well as the tanks of the receiving vessel, in order to ascertain the exact amount of fuel(s) received.

This is done to avoid malpractices creeping into the process of the fuel supply and to provide a certified statement of the quantity of FUEL OIL and DIESEL OIL finally delivered to the ship.

This type of survey will also involve,

(1.1) Sampling of the fuel supplied so that this may be analyzed to prove that what has been supplied is of accepted industry standards.

(1.2) All the calculations required in order to arrive at the quantities delivered onboard.

(1.3) Production of detailed bunker survey report(s).

(1.4) Sampling documentation and photographs as requested (if photography is permitted).

The most important aspect of a bunkering operation is the "checklist," which forms a part of the company's safety management system (SMS) and I.S.M., eliminating the possibility of human negligence and other operating errors.

The prebunkering checklist must be followed in-consultation with the Chief Engineer (C/E), as he is the person-in-charge for the bunkering operation.

Before bunkering the fourth Engineering Officer, takes "soundings" of bunker tanks and calculates the volume of fuel oil available in every fuel oil tank of the ship.

Then a Bunker-Plan will be made with regard to the distribution of the total quantity of bunker fuel oil to be received onboard.

PREBUNKERING PROCEDURE

1. State of adjacent waters noticed
2. Vessel properly secured to the dock

3. Check suppliers product corresponds to ordered product
4. Agree quantity to be supplied
5. Check valves open
6. Day tanks full and supply valves closed
7. Warning signs in position, for example, No Smoking
8. SOPEP plan available
9. Clean up material in place
10. Oil Boom in place
11. Foam fire extinguisher placed at bunker station
12. Transfer pumps off
13. Fuel tank supply valves open
14. Agree to stop/start signals between vessel and barge/truck
15. Bravo flag flying/red light showing
16. Agree the pumping/transfer rate
17. Agree emergency shut down procedure
18. Specification sheet received
19. Check hose and couplings are secure and in good order
20. Fuel nozzle and hose secured to the vessel
21. Check barge/truck meters reading
22. Check onboard meters reading
23. Bunker Valve open
24. Unused manifold connections blanked off
25. Master informed
26. Signal for the pumping to commence

The above checklist has to be completely and accurately filled by both the ship and the barge personnel.

SOPEP EQUIPMENT

At the bunker manifold and wherever necessary, as per the ships SOPEP plan, the SOPEP equipment should be kept in immediate readiness in order to avoid oil spills/pollution during any bunkering operation.

SOPEP: SHIPBOARD OIL POLLUTION EMERGENCY PLAN

The SOPEP Locker must have a minimum of the below specified items:

1. Absorbent roll
2. Absorbent pads
3. Absorbent granules
4. Absorbent materials
5. Brooms
6. Shovels
7. Mops

8. Scoops
9. Empty receptacles (200 L capacity)
10. Portable air driven pumps
11. Oil boom
12. Oil spill dispersants

These items must be stowed in an easily accessible locker, clearly marked, and are to be brought on deck ready for immediate use prior to all oil transfer operations.

DURING BUNKERING PROCEDURES

1. Witness taking and sealing of two representative product samples
2. Monitor fuel connections for leaks, fuel flow, and control tank levels
3. Change over of tanks whenever necessary
4. Checking the rate at which bunkers are received
5. Checking the tightness/slackness of mooring ropes
6. Checking the trim/list of the bunker barge and the ship
7. Continuous monitoring/lookouts for the vessel's position (when at anchor)

During bunkering, the above checklist must be filled in and continuous monitoring of the above specified items is required until the bunkering operation is complete.

THE BUNKER SURVEY IN DETAIL

The surveyor will ensure that:

– All necessary documentation is complete and available in his backpack prior to leaving his office to attend onboard the nominated vessel.
– All necessary materials and equipment are available for taking with him onboard.

   (1) All parties involved must be duly advised of the impending survey

     Master

     Shipowners

     Charterers

     Subcharterers (if applicable)

     All appointed Shipping Agents

     Any other attending surveyors, at the port where the survey is to be conducted

     The name of the attending surveyor(s) will be notified to all concerned.

   (2) The surveyor will ensure that the following manuals are available and ready for use:

PETROLEUM MEASUREMENT TABLES, COVERING

Volume Correction Factors

    Volume VIII

    Table 53B—Generalized Products

    Correction of Observed Density to Density at 15 °C

Table 54B—Generalized Products
Correction of Volume to 15 °C Against Density at 15 °C
ASTM-D-1250-80—API-ASTM-IP

WORKING WITH DENSITY AT 15 °C IN AIR

(1) Observed Ullage—apply corrections—get Corrected Ullage
(2) Observed Interface—apply corrections—get Corrected Interface
(3) From Corrected Ullage, find Total Observed Volume—*TOV (cubic meters)*
(4) From Corrected Interface, find Volume of Water—*WATER (cubic meters)*
(5) TOV-Water=Gross Observed Volume—*GOV of Cargo (cubic meters)*
(6) Use Density at 15 °C and Observed Temperature (C) and find Volume Correction Factor, from Table 54—*VCF*
(7) Gross Standard Volume=GOV×VCF—*(GSV) (cubic meters)*
(8) Weight Correction Factor=Density at 15° C in vacuum—0.0011 (This is the Density at 15° C in air.)—*WCF*
(9) Weight in Air=GSV×WCF (Density at 15° C in air)—*(Metric Ton)*

(10) Weight in Vaccum (Metric Ton)=GSV×Density at 15° C in vacuum

TOTAL OBSERVED VOLUME (TOV)

The total volume of material measured in the tank including

Cargo (oil or chemical)
Free water (FW)
Entrained sediment and water (S&W)
Sediment and scale
The foregoing should be as measured at observed temperature and pressure

FREE WATER (FW)

Water layer existing as a separate phase in the tanks, normally detected by water-paste or interface detector and usually settled at the bottom of the cargo tank depending on the relative density of the cargo.

SEDIMENT AND WATER (S&W OR BS&W)

Entrained material within the oil bulk, including solid particles and dispersed water, also sometimes known as base sediment and water (BS&W).

Expressed always as a percentage of the total cargo quantity is found out by collecting an average sample of the cargo inline during transfer and calculated by centrifuge technique in a laboratory.

GROSS OBSERVED VOLUME (GOV)

It is the total observed volume (TOV) less free water (FW) and bottom sediment, being the measured volume of product and sediment and water (S&W) at observed temperature and pressure.

Bottom sediment are normally not present on board a chemical or clean oil product tanker and therefore not deducted, whereas it may be present in a dirty oil carrier, but be very difficult to ascertain.

GROSS STANDARD VOLUME (GSV)

It is the measured volume of product and S&W at standard conditions of 15 °C and atmospheric pressure.

In practice, the GSV is the GOV multiplied by the volume correction factor VCF obtained from the appropriate ASTM/IP Petroleum Measurement Tables.

NET STANDARD VOLUME (NSV)

It is normally applied only to Crude Oils.

NSV is the GSV minus S&W, being a measurement of the *dry* oil quantity at standard conditions.

For clean oil products and chemicals, the S&W is not normally included within the receiver's quality specifications.

The term *Weight in Air* is that weight, which a quantity of fluid appears to have when weighed in air against standard commercial weights so that each will have a mass (weight in a vacuum) equal to the nominal mass associated with it.

The term *Weight in Vacuum* refers to the true mass of a fluid.

THE WEDGE FORMULA

The Wedge Formula is an approximation of the small quantities of liquid, solid cargo, and free water on board. The calculation can be carried out either before the vessel starts loading or after she has discharged.

The calculations take into account the dimensions of the individual cargo tank and the vessel's trim.

This formula is used when the oil liquid does not touch all bulkheads of the vessel's cargo tank.

The Wedge Formula assumes that:

1. Any "liquid" found in a cargo tank is in the form of a regular wedge shape with its base at the aft bulkhead of the cargo tank.
2. The shape of the wing tanks (the turn of the bilge) and in particular those wing tanks at the fore and aft parts of the vessel involves no curves (i.e., the tanks are box shaped) and that there are no appendices within the tanks, such as pipelines, or heating coils.
3. The trim of the vessel is small enough to accept that the "Sine" of its angle is the same as its "Tangent."
4. The ship has zero list and the liquid in the tank concerned is free flowing.

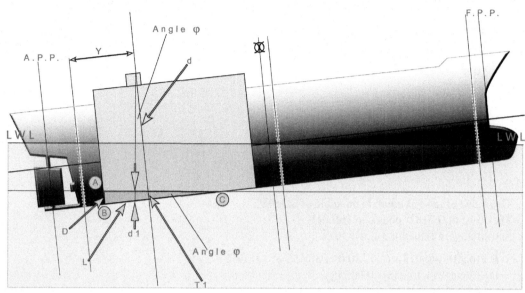

FIGURE 164

Wedge formula representation.

Let us consider a situation where a ship is trimmed and the liquid in a tank does not touch the forward bulkhead.

Take the sounding from the access-hatch so that the sounding tape bob is not restricted by a sounding pipe, that is, it is free to travel with the trim of the vessel or the barge. (Please refer Figure 164.)

The volume of the wedge ABC shown above is equal to V:

$$V = \frac{L \times B \times D^2}{2 \times T} \tag{1}$$

where:

V = Volume of liquid in the wedge
L = Length between perpendiculars (of the ship)
B = Breadth of the tank
D = Corrected dip
T = Ship's trim
$d1$ = Measured dip
Y = Distance from sounding position to after bulkhead
φ = Angle of trim
d = Depth of tank

$$D = \left[d1 \times \cosec\varphi + \left(Y - d \times \tan\varphi \right) \right] \times \tan\varphi \tag{2}$$

$$\frac{T}{L} = \tan\varphi$$

Example:
Consider a case where the following data applies:

$d1 = 20\,cm$
$Y = 2.0\,m$
$d = 30.0\,m$
$L = 360\,m$
$B = 20.0\,m$
$T = 2.0\,m$

$$\frac{2.0}{360.0} = \tan\varphi = 0.00555$$

This value of $\tan\varphi$ is equal to an angle of $0.3183°$.
The cosec of $0.3183°$ equals to 180.018.
Substituting in Equation (2), we have:

$D = [0.20 * 180.018 + (2.0 - 30.0 * 0.00555)] * 0.00555$
$D = [36.0036 + 1.8335] * 0.00555$
$D = 37.8331 * 0.00555$
$D = 0.21\,m$

Substituting in Equation (1), we have:

$V = [360.0 * 20.0 * 0.21^2]/2 * 2.0$
$V = 317.52/4$
$V = 79.38\,m^3$

DENSITY AND TEMPERATURE

These are important factors involved in ascertaining the weight of a fuel in vacuo and in air. If the declared value is not accurate, then errors will definitely be involved in the ensuing calculations.

Many a time, when we ask the Chief-Engineer to produce evidence of the fuel's density, we see a piece of paper containing handwritten data (including the density), which may even be unsigned by the issuing party. How reliable is this document?

Obviously, the attending surveyor has no alternative but to accept this figure. However, today there are portable instruments which can analyze a fuel sample and accurately produce its density. Equally, temperature sensors ought to be used to establish the temperature of the fuel in the tanks of the bunker barge.

The ship's command ought to carefully scrutinize the documents produced by the supplying barge, including the *Bunker Delivery Receipt*, customarily produced at the end of the bunkering operations. Watch out for what these documents state and if you are not satisfied with their contents, hand over to the Master of the barge a "Note of Protest."

It is not unheard of that the *Bunker Delivery Receipt* declares the volume delivered (as opposed to a weight); therefore, the ship's "Note of Protest" must counter all uncertainties.

The receiving ship's owners, via their Master, must reserve, in their "Note of Protest," their right to establish the weight of the fuel delivered, once an independent analytical laboratory has declared the fuel's density.

The aspect of temperature of the fuel delivered is not to be underestimated because it is well known and accepted that fuels are very susceptible to expansion and contraction in accordance with an increase and a decrease of their temperature, respectively.

As a consequence, it is important that the attending surveyor correctly and accurately records the fuel temperature at the time of the commencement of the pumping operations and again when nearing completion of the transfer.

The ship ought to have onboard apparatus capable of measuring both the density and the water content to calculate the mass of fuel delivered in her tanks. The cost of this equipment will be easily recovered by the ship's owners if there is a substantial dispute on bunkers.

Photos of the apparatus used for measuring the density and water content in fuel are shown below and on the next page.

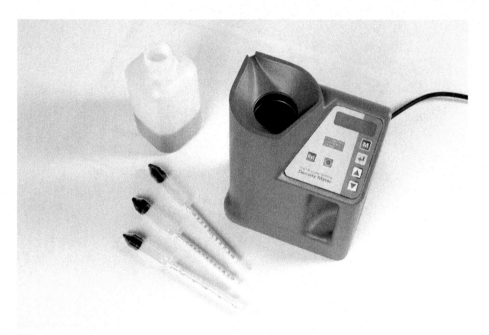

FIGURE 165

Fuel Oil density measuring apparatus.

Source: Parker Kittiwake.

The fuel that is delivered from the bunker barge is measured by a meter or by sounding. At this stage, it will be converted to mass by applying a density that has to be accurate in order to avoid errors creeping into the calculations. The shown apparatus is said to be as accurate as methods used in a laboratory (Figure 165).

The DIGI water in Oil test is said to provide a rapid indication of the total free and dissolved water content in an oil sample (Figure 166).

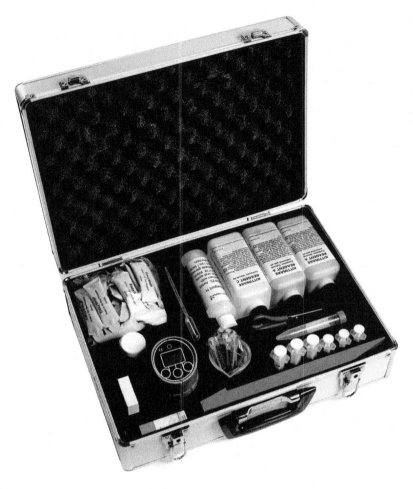

FIGURE 166

Fuel water content measuring apparatus.

Source: Parker Kittiwake.

FUEL AERATION

People in the trade call it the "Cappuccino Effect." Most liquids foam and froth when air is forcefully introduced into their mass. The reaction of marine fuel to such an introduction of a strong stream of air into it is to also foam and froth.

While this pressurized air does not adversely affect the properties of the fuel, it does cause it to expand. Once the fuel expands, any sounding obtained will be greater to that obtained with the fuel unaffected by external factors and calm.

The air is forced into the fuel by the supplying barge in order to "enhance" the soundings and as a result, the volume ends up being based on unrealistic soundings.

During the bunkering operations, the crew should monitor the quality of the fuel coming onboard. Every 30 min a sample should be obtained at the sampling point. If the oil is found to froth, board the barge and investigate the line blowing equipment.

Look for any tell-tale signs that the line blowing equipment was recently connected to the cargo line.

DILUTING THE TRUTH

The supplying barge may intentionally or unintentionally have fuel and water mixed before this fuel is delivered to the bunkering ship. Fresh water does not cause so much harm to the ship's machinery; however, salt water contains sodium, and if it finds its way into the combustion chamber, it will react with the vanadium in the fuel and cause corrosion.

The energy of fuel is reduced when water is mixed with the fuel. The disposal of water separated by the ship's plant (oily water separator, etc.) is a costly exercise. Further, these mixtures will damage other parts of the main-engine, such as exhaust valves and their seats and turbocharger blades.

A digital analysis kit that performs a "water in the oil" test is the best method for on-site and onboard testing. The use of the kit provides state of the art digital analysis and gives fast accurate results for easy monitoring of trends.

As a reminder of the ship's activities, we would recap as under:

The delivery of marine fuel is measured by volume, but paid for by mass. Accordingly, the fuel density must be established at the time of delivery. This is the way to implement good practice.

- The ship's personnel must ensure that there is a continuous drip method by which they investigate the whole process, from the beginning to the end.
- The ship's personnel must ensure that all fuel received onboard is placed in separate tanks so that this can be isolated if there are any problems with it.
- The ship's personnel must ensure that all of the documentation used by the supply-barge are genuine, free of corrections and alterations of dubious character and that they are approved and signed by the class of the vessel and/or by the local Maritime Authorities.
- The ship's personnel must ensure that they maintain an accurate record of any irregularities found or observed during the bunkering operations. This record will become an invaluable tool when it comes to issuing a "Note of Protest" to the supplying barge, if this is deemed necessary.

It is felt, at this point, that it is important to mention that Invensys UTC (part of theControl Group of the Department of Engineering Science at theUniversity of Oxford) has been in the forefront of developments in Coriolis Mass Flow meter, researching issues such as fast dynamic response and two-phase and multiphase flow. It is hoped that once this instrument enters commercial use a great number of the problems mentioned above will be solved.

So far, we have seen how a bunkering operation is executed when a vessel is to be supplied with fuel from a bunker barge.

There is another alternative and this involves the attendance of a surveyor onboard a ship for the purpose of sounding her tanks for the purpose of providing a certified statement of the quantity of:

(A) Heavy Viscosity Fuel Oil
(B) Marine Diesel Oil

This happens when the vessel enters a new charter or when the vessel is redelivered to her owners by the charterers.

The survey will be held onboard the vessel nominated by the attending surveyor's clients, and it will be conducted with the cooperation of the ship's command and in the presence of an officer nominated by the ship's Master or Chief-Engineer.

The attending surveyor will conduct an impartial and independent quantity measurement survey, regarding the marine fuels found onboard and he/she will issue a detailed bunker survey report to that effect. An agreement between the owners and the charterers will give the surveyor the right to sound or examine tanks that are not among those said to contain (A) or (B).

The report will include specific and detailed remarks on any bunker fuel shortages in order to enable the attending surveyor's principals to submit any fuel shortage claims, if they so desire.

The title of this section is directly related to the efforts of the parties which are determined to show that they have delivered the correct quantity of fuel to a third party when in fact they have supplied far less than required.

These efforts can, in some instances, be found to be relatively simple, but on occasion they can be quite intricate.

As a consequence, the attending surveyor must be aware (from the moment he starts with his/her attendance) that things may not be as they appear.

At the beginning of attendance and before any transfer of fuel—from a barge to a vessel—begins:

[1] The surveyor needs to measure all of the barge tanks that carry cargo, utilizing water finding paste.

[2] The surveyor needs to measure all of the barge tanks that carry her own fuel, utilizing water finding paste.

[3] The surveyor needs to check the tank calibration manuals to ensure that they are approved by an official body (class or government agency) and that the tables within have not been modified in any way.

[4] The surveyor needs to collect all relevant information concerning the fuel to be delivered to a ship, that is, the temperature prevailing in the cargo tanks at the time of his attendance and the density as per the document issued to the barge by the party that supplied the cargo of fuel. It is recommended that the surveyor asks for a photocopy of this document (if there is no photocopier onboard, the barge a photograph of this document ought to suffice).

[5] A set of samples ought to be collected from the tanks of the barge which are going to be used for the supply of fuel to the receiving vessel. These samples must be sealed in the presence of the surveyor and they must be labeled accordingly.

[6] The surveyor needs to record the trim and list of the barge.

[7] The Master of the barge is to be advised to watch out for the "inadvertent" introduction of any continuous air stream into the fuel that is being pumped onboard. Equally, the ship's officers are to keep the pumping operations under surveillance as the introduction of air into the fuel stream will cause the latter to foam and show a higher surface level (this adversely affecting the sounding) than what is actual.

[8] A running sample of the fuel supplied onboard ought to be obtained by the ship's officers.

[9] The ship's engineering officers ought to be asked to ensure that the fuel pumped onboard is kept in the predesignated tanks so that the calculations held on completion of the pumping operations will accurately reflect the total weight of the fuel supplied to the ship by the barge.

[10] The surveyor will sound the ship's fuel tanks, if these are not completely empty, applying trim and list corrections and using the last fuel receipt to establish the density of the fuel onboard.

[11] If on completion of the bunkering operations, there are still discrepancies in the quantities stemmed and the quantities delivered onboard steps [1] and [2] may have to be repeated.

[12] If the matter remains unresolved at the end of all efforts to pinpoint the cause of the discrepancy, the Master of the receiving vessel must hand over to the Master of the barge a suitable "Note of Protest."

It is important for a surveyor to attend onboard a ship before she starts bunkering operations and to follow the entire process, since he can serve as an independent witness in the event that there is a short delivery of fuel.

CASE 2: SURVEY OF THE FUEL EXISTING IN THE TANKS OF AN OCEAN TRADING VESSEL

What we dealt with in the foregoing text can be reversed when the party receiving a quantity of fuel is determined to show that they have received a lesser quantity of fuel by a third party when in fact they have received what they paid for.

In order for the surveyor to ensure that there are no peculiarities in the conducting of his survey he should adopt the following (Figure 167):

[1] Before starting with taking the sounding of the tanks, the surveyor must ascertain that the tape used is accurate.

For this purpose, he should check the length of the tape against his measuring tape.

Checking the first 30 cm, including the bob, ought to be sufficient.

[2] On sounding the ship's fuel oil tanks, the surveyor ought to measure the total height of the pipe once the sounding bob has hit the striking plate.

It is not always the case that sounding pipes on the port side are of equal height to those sounding pipes on the starboard side, however, in most cases they are.

If in doubt, there are drawings onboard that show the height of each pipe.

[3] Adopting a particular format while recording the soundings, the surveyor may put himself in a better position to judge an otherwise difficult situation.

Once the surveyor has completed his attendance on the deck and in the engine-room collecting soundings and other required information, he ought to consult the ship's calibration books and record the maximum sounding height and the corresponding tank volume in his documents, for later reference purposes (if required).

[4] If in due course the surveyor finds that the height of pipes symmetrically located onboard differs considerably, he ought to investigate matters further.

It is not impossible that the difference is because a rod or part of a sounding tape may have fallen to the very bottom of the sounding pipe, blocking part of it.

This situation will introduce errors as it will not be possible to obtain an accurate sounding. The only remaining alternative (if possible) will be to obtain an ullage instead of a sounding.

An example of the data to be recorded in the surveyor's notebook is shown in p. 207. A record of the sounding, the pipe height, the water trace (if any), and all pertinent notes are to be entered. If any additional notes are required, these can be appended on a separate page (Figures 168 and 169).

Rumor has it that sometimes, for obscure reasons, the officers or the crew of a vessel would wish to show a different sounding to that actually corresponding to a particular tank. The methods used to achieve that purpose are shown in p. 208.

FIGURE 167

Verifying the length of a sounding tape to a measuring tape.

M/V'...,

Draft Aft:	Draft Amid. P:	Survey Date: 2015
For'd:	Draft Amid. S:	Survey Soundings Start: ....hrs.
Trim:	List:	Survey Soundings Finish: ....hrs.

Bunker Survey					Notes/Remarks
Tank no.	Sounding	Height	Water		

FIGURE 168

Sample page of the surveyor's notebook.

FIGURE 169

Tank sounding pipes on weather deck.

[5] Pipe arrangement no.1

In this instance, a blind flange is placed, say at the first flange connection of the sounding pipe.

Those involved have then poured 4 or 5 cm worth of oil in the pipe. This amount will result in a 5 cm sounding, whereas the tank contains a far greater amount of oil and would—if it had been properly sounded—have produced a far greater sounding.

This way the tank, via this modified sounding, appears to contain a smaller quantity of fuel than it actually does. However, if one were to pour a bucket of oil in this arrangement (in the sounding pipe), this relatively small amount of oil would make the sounding pipe overflow. Obviously, this is something that would happen only if the tank was completely full, which is not the case (see Figure 170 LHS).

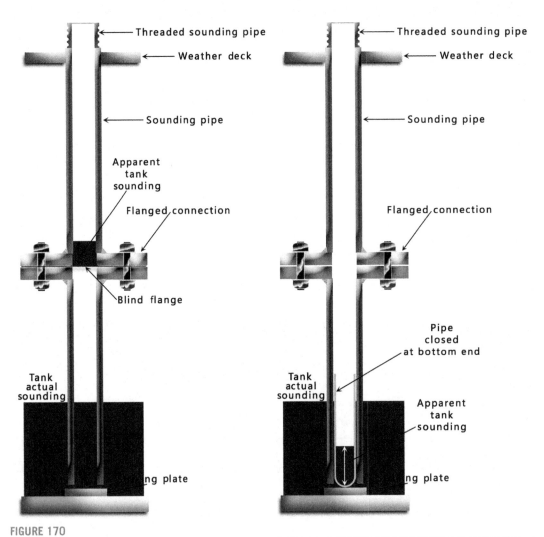

FIGURE 170

LHS photo—Pipe arrangement no. 1. RHS photo Pipe arrangement no. 2.

[6] Pipe arrangement no.2.

An alternative arrangement is that shown in "Figure 170 RHS." In this instance, a pipe which fits snugly within the sounding pipe has been blanked-off at one end and it has been filled with a small quantity of the same oil as in the tank to which the sounding pipe corresponds.

The pipe has then been lowered until it comes to rest on the striking plate of the tank. Once again, if the tank is dipped it will produce a sounding which is substantially smaller than the sounding of the actual fuel in the tank.

The above are relatively simple devices which can be used to alter things to someone's advantage. However, on some occasions, rumor has it that enterprising individuals have erected false bulkheads within cargo tanks with a top overflow opening. This means that once the oil level reaches that opening the oil will flow to the other side of the bulkhead never to be seen again.

BUNKER SURVEY DOCUMENTATION

As with every survey, the production of a satisfactory final report will be as good as the notes one has collected during the attendance.

We have already seen the outline of the surveyors notebook (please refer Figure 168).

The following data are given for guidance purposes, and it reflects what would apply to a 44,000.00 tonne bulk carrier.

Cargo Hold No.		Capacity (Tonnes)
1..		8946.00
2..		13,090.00
3..		10,490.00
4..		13,092.00
5..		9668.00
Heavy Fuel Oil Tanks		**Capacity (Cubic Meters)**
Storage	1S	292.00
Storage	2P	199.00
Storage	2S	373.00
Storage	5P	163.00
Storage	5S	163.00
Service	SERVICE1	32.00
Service	SERVICE2	39.00
Settling	SETTLING1	36.00
Settling	SETTLING2	40.00
Diesel Oil Tanks		**Capacity (Cubic Meters)**
Storage	1P	67.00
Storage	2P	40.00
Service	SERVICE1P	16.00
Service	SERVICE2P	16.00

Now let us look at how a heel and trim correction is carried out.

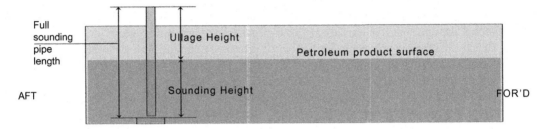

In respect of the tank shown above, we have:

TRIM = Draft at A.P.P.—Draft at F.P.P

TRIM = expressed in meters

HEEL = [(Port Draft at Amidships—Stbd Draft at Amidships)/Ship's Beam] * 57.3

HEEL = expressed in degrees

For our particular ship, the heel correction table is as below:

				Heel Correction Table (cm)				
Sounding (m)	Heel 2.0°(P)	Heel 1.5°(P)	Heel 1.0°(P)	Heel 0.5°(P)	Heel 0.5°(S)	Heel 1.0°(S)	Heel 1.5°(S)	Heel 2.0°(S)
1.00	4.8	3.0	1.9	0.3	3.7	9.1	15.0	21.4
1.10	−0.2	−1.4	−2.3	−2.2	3.5	8.1	13.4	19.5
1.20	−4.5	−4.9	−5.0	−3.1	3.8	8.1	13.2	18.8
1.30	−7.8	−7.4	−6.1	−3.4	4.0	8.6	13.6	19.1
1.40	−10.2	−9.1	−6.8	−3.7	4.3	9.1	14.2	19.8
1.50	−12.1	−10.3	−7.4	−4.0	4.6	9.4	14.8	20.5
1.60	−13.6	−11.1	−7.9	−4.2	4.8	9.8	15.3	21.1
1.70	−14.8	−11.8	−8.4	−4.4	4.9	10.1	15.7	21.7
1.80	−15.8	−12.5	−8.8	−4.6	5.0	10.4	16.2	22.3
1.90	−16.6	−13.1	−9.2	−4.8	5.1	10.7	16.6	22.8

The trim correction table is as under:

				Trim Correction Table (m³)				
Sounding (m)	Even Keel	Trim (−2.0 m)	Trim (−1.0 m)	Trim (1.0 m)	Trim (2.0 m)	Trim (3.0 m)	Trim (4.0 m)	Trim (5.0 m)
1.60	6.6	15.7	1.7	0.8	0.4	0.5	0.5	0.5
1.62	11.5	20.8	5.7	2.2	1.7	1.2	0.8	0.8
1.64	16.3	25.9	9.7	5.0	3.0	2.5	2.0	1.5
1.66	21.2	31.1	13.7	8.0	5.0	3.9	3.3	2.7
1.68	26.1	36.2	17.7	10.9	7.3	5.5	4.5	3.8
1.70	31.3	41.3	22.4	14.7	10.9	8.3	6.3	5.2

Trim Correction Table (m³)								
Sounding (m)	Even Keel	Trim (−2.0 m)	Trim (−1.0 m)	Trim (1.0 m)	Trim (2.0 m)	Trim (3.0 m)	Trim (4.0 m)	Trim (5.0 m)
1.72	36.5	46.5	27.5	19.5	14.6	11.4	9.0	6.9
1.74	41.8	51.9	32.6	24.3	18.4	14.5	11.7	9.4
1.76	47.1	57.4	37.7	29.1	22.1	17.5	14.3	11.9
1.78	52.3	62.8	42.8	33.9	25.8	20.6	17.0	14.3
1.80	62.0	72.0	52.0	43.9	35.0	30.0	27.0	24.0

Important note: (−) sign signifies trim by the head. No signal signifies trim by the stern.

Example:

During a bunker survey, the motor-vessel "ALBION 35" is found to be trimming 1.35 m by the stern. She is also heeling 1.20° to starboard. No water trace was found in the petroleum product.

Tank no. 2 (S) is found with a sounding of 1.675 m.

Find the volume of the petroleum product in this tank.

Consulting the corrections given in the above two tables and applying linear interpolations, we have as follows:

	Heel Correction (cm)		
	1.0°(S)	1.2°(S)	1.5°(S)
Sounding			
1.60 m	9.8		15.3
1.675 m	10.025	12.255	15.600
1.70 m	10.1		15.7

Therefore, heel correction is 1.675 m + 0.122 m = 1.797 m.

	Trim Correction (m³)		
	1.0 m	1.35 m	2.0 m
Sounding			
1.780 m	33.9		25.8
1.797 m	42.4	39.327	33.62
1.800 m	43.9		35.0

The volume of the petroleum product in tank no. *2 (S)* is found to be *39.327 m³*.

For reference only, an example of the complete tabulation of a bunker survey follows below (this usually includes Fuel Oil and Diesel Oil). The entries (except for the tabulations and their results) are shown to reflect a realistic situation but are presented for reference only.

Bunker Fuel Measurement Report

Petroleum Measurement Tables utilized: ASTM D 1250-80 Table 54B and volume XII, Table 21

Ship's Name:	**ALIBION 35**				
Port of Survey:	PORT RASHID, DUBAI, U.A.E.				
Survey Date:	**27.01.2012**	**Time: Start:**	**0830 hours**	**Finish:**	**1100 hours**
Draft Aft: 6.00 m Draft For'd: 5.00 m Trim: 1.00 m		Draft Amidships, Port: Draft Amidships, Stbd: Heel:		5.20 5.20 Nil	

The receipt furnished by the ship's Chief-Engineer for this fuel stated the following:

API Gravity at 60 F	Relative Density 60/60 F	Density at 15 C
-----	-----	0.9706 and 0.9780
Sea Condition:	Calm	

Product surveyed	**FUEL OIL**							
Tank Designation	**Sounding Obs.**	**Sounding Corr.**	**Temp. Obs.**	**Density 15/15 C**	**Volume Trim Corr.**	**Density in Air**	**Volume Corr. Factor**	**Product Weight in Air**
TANK 23 (S)	7.59		36	0.9706	246.00	0.9695	0.9852	234.970
TANK 27 (S)	7.93		36	0.9780	297.31	0.9769	0.9854	286.201
TANK 28 (P)	8.04		36	0.9780	303.49	0.9769	0.9854	292.149
TANK 24 (P)	7.89		36	0.9706	258.81	0.9695	0.9852	247.206
TANK 20 (P)	0.07		36	0.9706	1.07	0.9695	0.9852	1.021
TANK 26 (P)	6.73		36	0.9780	188.80	0.9769	0.9854	181.750
TANK 25 (S)	6.60		36	0.9780	166.92	0.9769	0.9854	160.683
TANK 52			75	0.9706	15.00	0.9695	0.9576	13.926
TANK 48			45	0.9706	48.00	0.9695	0.9789	45.554
TANK 47			42	0.9706	30.00	0.9695	0.9810	28.532
TOTAL FUEL OIL IN ALL TANKS SOUNDED, METRIC TONS, IN AIR								**1491.992**

The receipt furnished by the ship's Chief-Engineer for this fuel stated the following:

API Gravity at 60 F	Relative Density 60/60 F	Density at 15 C
-----	-----	0.8704

Product surveyed	**DIESEL OIL**							
Tank Designation	**Sounding Obs.**	**Sounding Corr.**	**Temp. Obs.**	**Density 15/15 C**	**Volume Trim Corr.**	**Density in Air**	**Volume Corr. Factor**	**Product Weight in Air**
TANK PORT	1.93		28	0.8704	33.05	0.8693	0.9895	28.429

TANK STBD	4.04		28	0.8704	50.06	0.8693	0.9895	43.060
TANK SERVICE	2.53		26	0.8704	5.43	0.8693	0.9911	4.678
TOTAL DIESEL OIL IN ALL TANKS SOUNDED, METRIC TONS, IN AIR								**76.167**

Once all of the soundings have been obtained, the weight calculations will follow, which include all trim and heel corrections as well as the introduction of volume correction factors and density in air. These data will be found in the ASTM tables, but there are ways for those who can do coding, to create a small piece of software which with the input of trim corrected volume, density, and temperature, can rapidly calculate the weight of fuel oil or diesel oil in air. Alternatively, MS Excel can also be used. You may start by using linear interpolations to introduce corrections for trim, heel, and trim corrected volume. If your knowledge of this spreadsheet can take you further, you can calculate volume correction factors and add the last three columns (see example on the previous pages) to complete the process.

Always ask the Chief-Engineer whether all of the fuel tanks contain a petroleum product of one density. In some instances, the vessel has replenished bunkers and two fuels of different densities have been mixed in the same tank.

If the Chief-Engineer confirms that this is the case, an adjustment must be made in order to arrive at an *equivalent density* of the mixture of petroleum products. This adjustment needs to be applied to each tank that contains a mixture of fuels.

Example:

A ship's double-bottom tank (no. 3) contains a volume of $567.00\,m^3$ of Fuel Oil of a density of 0.9956 at 15°/15 °C.

The vessel arrives at the next port where she replenishes $250.00\,m^3$ of Fuel Oil of a density of 0.9600 at 15°/15 °C. This fuel for ship's convenience is mixed with the fuel already existing in double bottom tank no. 3.

Find the new density or equivalent density of the fuel mixture.

$$[567.00 * 0.9956] + [250.00 * 0.9600] = 817.00 * D,$$

Where is the equivalent density of the fuel mixture in the double-bottom tank no. 3.

$$564.505 + 240.000 = 817.00 * D$$

$$D = 804.505 / 817.00 = 0.9847 \text{ at } 15° / 15°C.$$

Below follows an example of the certificate to be issued on completion of the bunker survey.

FUEL MEASUREMENT CERTIFICATE
TABLES USED:ASTM D 1250 80

SHIP'S NAME:			
SHIP'S FLAG			
SHIP'S PORT OF REGISTRY			
SHIP'S OFFICIAL REGISTRY NUMBER			
BUNKER TANKS SOUNDED ON / AT	DATE:		TIME:

SURVEY ON DELIVERY:/ SURVEY ON RE-DELIVERY:
(Delete whichever is inapplicable),
SURVEY ON:_____
(If this survey has not been held for Delivery or Redelivery purposes state the occasion on which it was held)

THIS IS TO CERTIFY that the undersigned surveyor, did attend onboard the vessel whose particulars are shown above for the purpose of holding survey of the quantity of bunkers remaining onboard upon:
Delivery / Redelivery / Other occasion, state
particulars_____
_____

The Master advised that Delivery took place / Redelivery will take place on the date and time shown in the box below.
This survey has been held in accordance with instructions received from:
_____
This survey was held whilst the said vessel lay afloat alongside berth no.:_____
In the port of: _____

All tanks presented to contain bunkers were sounded along with one of the ship's officers. The ship's calibration tables have been utilized in these calculations and after applying the necessary corrections the following quantities of fuel oils were ascertained to remain onboard at the time of Delivery / Redelivery / (Other occasion, state particulars)
_____

(Delete whichever is inapplicable).

HEAVY VISCOSITY FUEL: _____Metric Tons
MARINE DIESEL OIL:_____Metric Tons
DELIVERY / REDELIVERY / Other occasion, as stated above.
DATE:_____ TIME _____hours
(Delete whichever is inapplicable)

Master   **Chief-engineer**   **Surveyor**

NB: The ship's stamp must be placed next to the Master's signature.

NOTE:11-253 ----- In the event that delivery or re-delivery takes place before or after holding a bunker survey, the attending surveyor must make an adjustment on the basis of the elapsed time and the specific Fuel oil and Diesel oil consumptions stated in the Charter Party in force. The explanatory notes to this effect must be included in the surveyor's formal report.

SURVEY COMPLETION

FINAL CONSIDERATIONS UPON COMPLETION OF SURVEY

Once you have completed your attendance onboard a ship with respect to an "On-Hire," or an "Off-Hire" survey, it is time to collect your thoughts and review all of your notes, including the documentation that you might have collected while holding the bunker survey and/or the condition survey.

This is the time to revisit any forms or questionnaires that you have brought onboard among your other documents. The surveys you have completed require arduous work and in a way gruelling work, so it is not impossible that you may have forgotten to ask a certain question or to have collected a piece of data that you intended to take with you for reference while writing your formal report.

Now is the time to reflect and spend a little longer getting your thoughts together. You have spent hours making sure that you did your work diligently, so a few extra minutes might bring forward some aspects you forgot to attend to. Maybe a last look on deck will remind you of something you forgot. Similarly, think about whether you have asked all the questions that you intended to ask. There is also another point to consider. Did you see something onboard this ship that you had not come across in the past and you would wish to add it to your standard questionnaire, for future use? Now is the time to make a note.

Sometimes you may say that you will make a mental note to write down something later and when "later" comes you have forgotten what that something was. Do not leave for later, something that can be done on the spot, persist and you will thank yourself for sticking with a good practice.

COORDINATION WITH THE SHIP'S COMMAND

On boarding a ship, the surveyor should make his/her first priority to meet with the ship's Master, introduce himself to him, and state the purpose of his being onboard.

The surveyor should explain how he is proposing to carry out his inspection and discuss with the ship's Master any particular aspects he wishes to cover.

He should also advise the Master as to how he intends to round up his attendance onboard and bring the survey to an end.

At this point, one of the ship's officers should be allocated to accompany the surveyor during his attendance.

This is good practice, but it is also a process necessary to take care of safety aspects.

One of the important points here is the issuing of any recommendations to the vessel.

On occasion, these recommendations may cover deficiencies which are severe enough to require that the ship must have such deficiencies satisfactorily rectified before the ship proceeds to sea on her next employment.

In such cases, it is ideal that repairs are effected and the surveyor is given the opportunity to inspect the repaired item(s) so that he may delete the recommendation(s) given in the first instance.

THE DOCUMENTATION TO BE COMPLETED PRIOR TO LEAVING THE VESSEL

The surveyor, on completion of his attendance, will issue an:

[ON-HIRE, or DELIVERY Certificate] or [OFF-HIRE or REDELIVERY Certificate] as the case may be. A sample is shown below:

ON-HIRE / DELIVERY CERTIFICATE

THIS IS TO CERTIFY

That the undersigned surveyor, did attend onboard the vessel whose particulars are shown below, for the purpose of holding a survey upon her Delivery to:

_____ [enter the Charterers' name]

SHIP'S NAME:_____
SHIP'S FLAG:_____
SHIP'S PORT OF REGISTRY:_____
SHIP'S OFFICIAL REGISTRY NUMBER:_____

The Master advised that the vessel went ON-HIRE on the _____ at _____ hours.
This survey has been held in accordance with instructions received from:

_____
_____

This survey was held whilst the said vessel lay afloat alongside berth no.:_____in the port
of:_____

All tanks presented to contain bunkers were sounded along with one of the ship's officers. The ship's calibration tables have been utilized in these calculations and after applying the necessary corrections, the following quantities of fuel oils were ascertained to remain onboard at the time of Delivery:
FUEL OIL: _____Metric Tons
DIESEL OIL:_____Metric Tons

During the ON-HIRE CONDITION survey a total of _____ defects were recorded and a list of these defects, together with respective recommendations, was handed over to the ship's Master for his consideration and action.

Master   **Chief-Engineer**

                              AttendingSurveyor

The ship's stamp

                                        Signed on_____
must be placed

                                        At:_____hours.
over the Master's
signature.

SURVEY REPORT: A PROPOSED FORMAT

In the following pages, the reader will find a proposed *ON-HIRE/OFF-HIRE* report format. This proposal has been prepared to offer a guideline as to the various data that ought to be collected and the findings to be recorded upon survey, in respect of the ship's condition.

In the event that the attending surveyor is conducting an OFF-HIRE survey, he/she ought to consult the ON-HIRE report so that it can be determined whether new damage has occurred, for which the Charterers may be liable to arrange reinstatement to a satisfactory condition or to compensate the Owners (Figure 171).

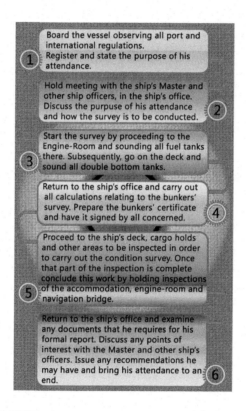

1. Board the vessel observing all port and international regulations. Register and state the purpose of his attendance.

2. Hold meeting with the ship's Master and other ship officers, in the ship's office. Discuss the purpose of his attendance and how the survey is to be conducted.

3. Start the survey by proceeding to the Engine-Room and sounding all fuel tanks there. Subsequently, go on the deck and sound all double bottom tanks.

4. Return to the ship's office and carry out all calculations relating to the bunkers' survey. Prepare the bunkers' certificate and have it signed by all concerned.

5. Proceed to the ship's deck, cargo holds and other areas to be inspected in order to carry out the condition survey. Once that part of the inspection is complete conclude this work by holding inspections of the accommodation, engine-room and navigation bridge.

6. Return to the ship's office and examine any documents that he requires for his formal report. Discuss any points of interest with the Master and other ship's officers. Issue any recommendations he may have and bring his attendance to an end.

FIGURE 171

A synopsis of the attending surveyor's activities.

The report comprises four (4) parts, as follows:
Part 1. SURVEY REPORT PREAMBLE,
Part 2. SHIP CONDITION SURVEY,
Part 3. SHIP DEFECTS' LIST,
Part 4. SHIP BUNKERS' SURVEY FINDINGS.

SURVEY REPORT PREAMBLE (PART ONE)

Report Reference no:-   COMMERCIAL SHIP SURVEYING/ON-HIRE SURVEY REPORT/54321
Report Date:-   1st May 2015.

THIS IS TO CERTIFY

that I,

[Insert the name of the attending surveyor]

have been instructed by:-

[Insert the name of the party instructing the surveyor]

to hold surveys, upon their
having received a request
for the appointment of a surveyor from:-

[Insert the name of the party requesting the survey]

The purpose of this attendance has been:-

[Insert the full details of the survey carried out]

Surveys held onboard the captioned vessel were carried out with the consent of her Master. Particulars, findings, and details relevant to this case follow within this report.

Ship's Particulars.

Survey application date	
Attendance commenced	
Attendance completed	
Report date	

[Insert the dates relevant to this survey]

Name	
Former Name	
IMO Number	
Flag	
Port of Registry	
Official Registry Number	
Call Sign	

[Insert the ship's particulars, group 1]

Gross Tonnage	
Net Tonnage	
Displacement, maximum	
Deadweight, summer marks	

[Insert the ship's particulars, group 2]

Length, overall	
Breadth, molded	
Draft, summer	
Depth, molded	

[Insert the ship's particulars, group 3]

Shipbuilders	
Year of Build	
Main Engine, Makers	
Main Engine, Type	
Main Engine, Power	
Classification Society	

[Insert the ship's particulars, group 4]

[002.06:Insert the ship's particulars, group 5]

Ship's Owners	

[Insert the ship owners' particulars, group 6]

Ship's Operators	

[Insert the ship operators' particulars, group 7]

Ship Officers' Particulars.

Master	
	[Insert the Master's full name, group 8]
Certificate of Competency	
	[Insert the CoC class, group 9]
CoC endorsements (if any)	
	[Insert endorsements in the CoC class, group 10]
Years (1) at sea (2) in rank	years	years
	[Insert number of years in respective boxes, group 11]
Chief-Engineer	
	[Insert the Chief-Engineer's full name, group 11]
Certificate of Competency	
	[Insert the CoC class, group 9]
CoC endorsements (if any)	
	[Insert endorsements in the CoC class, group 10]
Years (1) at sea (2) in rank	years	years
	[Insert number of years in respective boxes, group 11]

Ship Brief Description.

Motor-Vessel	
	[Insert the ship's name]

Construction	All welded steel construction
	Partly riveted and partly welded construction
	[Delete whichever does not apply]

General configuration	Bulk carrier
	Multi purpose general-cargo carrier
	Container ship
	Ro-Ro
	Reefer
	[Delete whichever does not apply]

Number of cargo holds		Cargo holds for'd of the accommodation
		Cargo holds aft of the accommodation
		[Insert the number of cargo holds in respective boxes]

Hatches, per each hold	Hold no. 1	Single	Tween	Hatch(es)
	Hold no. 2	Single	Tween	Hatch(es)
	Hold no. 3	Single	Tween	Hatch(es)
	Hold no. 4	Single	Tween	Hatch(es)
				[Delete inapplicable types of hatches and cargo holds]

Hatch covers makers and type	Hold no. 1	
	Hold no. 2	
	Hold no. 3	
	Hold no. 4	
		[Insert makers and type of covers, in each row]

Tween deck hatch covers, type	Hold no. 1	
	Hold no. 2	
	Hold no. 3	
	Hold no. 4	
		[Insert the type of tween deck covers, in each row]

Cargo hold ventilation, type	Hold no. 1	
	Hold no. 2	
	Hold no. 3	
	Hold no. 4	
		[Insert the type of ventilation, delivery and suction, in each row]

Cargo gear, per cargo hold	Hold no. 1	
	Hold no. 2	
	Hold no. 3	
	Hold no. 4	
		[Insert the type of cargo gear and its SWL, in each row]

Cargo battens and tanktop sheathing	Hold no. 1	
	Hold no. 2	
	Hold no. 3	
	Hold no. 4	
		[Insert notes and remarks in each row]

Ship Port Rotation.

The information provided by the ship's Master regarding her ON-HIRE, her recent port rotation and her OFF-HIRE data has been as shown below

ON-HIRE

Port	
Call purpose	
Arrival date/time	
Departure date/time	

RECENT PORTS OF ALL

Port	
Call purpose	
Arrival date/time	
Departure date/time	

Port	
Call purpose	
Arrival date/time	
Departure date/time	

Port	
Call purpose	
Arrival date/time	
Departure date/time	

Port	
Call purpose	
Arrival date/time	
Departure date/time	

OFF-HIRE

Port	
Departure date/time	

[Insert port name and dates as mm/dd/yyyy, time as hh:mm]

SHIP CONDITION SURVEY (PART TWO)
SURVEY OF HULL

ITEMS INSPECTED	Struct./Mech.	Coating
1.		
Port side bow shell plating		
2.		
Port side shell plating		
3.		
Port side stern shell plating		
REMARKS:		

ITEMS INSPECTED	Struct./Mech.	Coating
4.		
Stbd side bow shell plating		
5.		
Stbd side shell plating		
6.		
Stbd side stern shell plating		
REMARKS:		

ITEMS INSPECTED	Struct./Mech.	Coating
7.		
Port side boottopping		
8.		
Stbd side boottopping		
9.		
Transom stern shell plating		
REMARKS:		

ITEMS INSPECTED	Struct./Mech.	Coating
10.		
Bulbous bow plating		
11.		
Port side anchor, chain stoppers, chains.		
12.		
Stbd side anchor, chain stoppers, chains.		
REMARKS:		

ITEMS INSPECTED	Struct./Mech.	Coating
13.		
Aft port side draft marks		
14.		
Aft Stbd side draft marks		
15.		

REMARKS:		

ITEMS INSPECTED	Struct./Mech.	Coating
16.		
Amidships Port side draft marks		
17.		
Amidships Stbd side draft marks		
18.		

REMARKS:		

ITEMS INSPECTED	Struct./Mech.	Coating
19.		
For'd Port side draft marks		
20.		
For'd Stbd side draft marks		
21.		

REMARKS:		

ITEMS INSPECTED	Struct./Mech.	Coating
22.		
Port side gangway		
23.		
Stbd side gangway		
24.		

REMARKS:		

SURVEY OF WEATHER DECK AND CROSSDECKS

ITEMS INSPECTED	Struct./Mech.	Coating
25.		
Port side poop deck shell plating, incl. bollards and fairleads		
26.		
Stbd side poop deck shell plating, incl. bollards and fairleads		
27.		
Stern deck shell plating, incl. equipment and mooring winches		
REMARKS:		

ITEMS INSPECTED	Struct./Mech.	Coating
28.		
Port side weather deck shell plating		
29.		
Port side bulwark and railings		
30.		
Fo'c'sle blkhd and crossdeck structure, outfit incl. spare anchor		
REMARKS:		

ITEMS INSPECTED	Struct./Mech.	Coating
31.		
Fo'c'sle plating, bollards, fairleads, bulwarks and mast		
32.		
Fo'c'sle anchor windlass port and stbd incl. accessories		
33.		
Fo'c'sle mooring winches port and stbd incl. accessories		
REMARKS:		

ITEMS INSPECTED	Struct./Mech.	Coating
34.		
Bossun's stores incl. accessories and equipment		
35.		
Chain locker incl. accessories and equipment		
36.		
Fore peak tank (internal inspection)		
REMARKS:		

ITEMS INSPECTED	Struct./Mech.	Coating
37.		
Emergency fire pump		
38.		

39.		

REMARKS:		

ITEMS INSPECTED	Struct./Mech.	Coating
40.		
Hatch no. 1 – For'd hatch coaming structure, outfit, equipment		
41.		
Hatch no. 1 Port hatch coaming structure, outfit, equipment		
42.		
Hatch no. 1 Aft hatch coaming structure, outfit, equipment		
REMARKS:		

ITEMS INSPECTED	Struct./Mech.	Coating
43.		
Hatch no. 1 stbd hatch coaming structure, outfit and equipment		
44.		

45.		

REMARKS:		

ITEMS INSPECTED	Struct./Mech.	Coating
46.		
Hatch cover no. 1 structure, outfit and equipment		
47.		

48.		

REMARKS:		

ITEMS INSPECTED	Struct./Mech.	Coating
49.		
Hatch no. 2 – For'd hatch coaming structure, outfit, equipment		
50.		
Hatch no. 2 Port hatch coaming structure, outfit, equipment		
51.		
Hatch no. 2 Aft hatch coaming structure, outfit, equipment		
REMARKS:		

ITEMS INSPECTED	Struct./Mech.	Coating
52.		
Hatch no. 2 stbd hatch coaming structure, outfit and equipment		
53.		

54.		

REMARKS:		

ITEMS INSPECTED	Struct./Mech.	Coating
55.		
Hatch cover no. 2 structure, outfit and equipment		
56.		

57.		

REMARKS:		

ITEMS INSPECTED	Struct./Mech.	Coating
58.		
Hatch no. 3 – For'd hatch coaming structure, outfit, equipment		
59.		
Hatch no. 3 Port hatch coaming structure, outfit, equipment		
60.		
Hatch no. 3 Aft hatch coaming structure, outfit, equipment		
REMARKS:		

ITEMS INSPECTED	Struct./Mech.	Coating
61.		
Hatch no. 3 Stbd hatch coaming structure, outfit and equipment		
62.		

63.		

REMARKS:		

ITEMS INSPECTED	Struct./Mech.	Coating
64.		
Hatch cover no. 3 structure, outfit and equipment		
65.		

66.		

REMARKS:		

ITEMS INSPECTED	Struct./Mech.	Coating
67.		
Hatch no. 4 – For'd hatch coaming structure, outfit, equipment		
68.		
Hatch no. 4 Port hatch coaming structure, outfit, equipment		
69.		
Hatch no. 4 Aft hatch coaming structure, outfit, equipment		
REMARKS:		

ITEMS INSPECTED	Struct./Mech.	Coating
70.		
Hatch no. 4 Stbd hatch coaming structure, outfit and equipment		
71.		

72.		

REMARKS:		

ITEMS INSPECTED	Struct./Mech.	Coating
73.		
Hatch cover no. 4 structure, outfit and equipment		
74.		

75.		

REMARKS:		

ITEMS INSPECTED	Struct./Mech.	Coating
76.		
Cross deck – Fo'c'sle blkhd / hatch no. 1 – structure, outfit, etc.		
77.		
Cross deck – Hatch no. 1 / hatch no. 2 – structure, outfit, etc.		
78.		
Cross deck – Hatch no. 2 / hatch no. 3 – structure, outfit, etc.		
REMARKS:		

ITEMS INSPECTED	Struct./Mech.	Coating
79.		
Cross deck – Hatch no. 3 / hatch no. 4 – structure, outfit, etc.		
80.		
Cross deck – Hatch no. 4 / Accom. blkhd – structure, outfit, etc.		
81.		

REMARKS:		

SURVEY OF MASTHOUSES AND CARGO GEAR

ITEMS INSPECTED	Struct./Mech.	Coating
82.		
Masthouse no. 1		
83.		
Crane no. 1		
84.		

REMARKS:		

ITEMS INSPECTED	Struct./Mech.	Coating
85.		
Masthouse no. 2		
86.		
Crane no. 2		
87.		

REMARKS:		

ITEMS INSPECTED	Struct./Mech.	Coating
88.		
Masthouse no. 3		
89.		
Crane no. 3		
90.		

REMARKS:		

ITEMS INSPECTED	Struct./Mech.	Coating
91.		
Masthouse no. 4		
92.		
Crane no. 4		
93.		

REMARKS:		

ITEMS INSPECTED	Struct./Mech.	Coating
94.		
Domestic crane		
95.		

96.		

REMARKS:		

SURVEY OF TOPSIDE TANKS: PORT (INTERNAL INSPECTION)

ITEMS INSPECTED	Struct./Mech.	Coating
97.		
Port topside tank no. 1 (internal inspection)		
98.		
Stbd topside tank no. 1 (internal inspection)		
99.		

REMARKS:		

ITEMS INSPECTED	Struct./Mech.	Coating
100.		
Port topside tank no. 2 (internal inspection)		
101.		
Stbd topside tank no. 2 (internal inspection)		
102.		

REMARKS:		

ITEMS INSPECTED	Struct./Mech.	Coating
103.		
Port topside tank no. 3 (internal inspection)		
104.		
Stbd topside tank no. 3 (internal inspection)		
105.		

REMARKS:		

ITEMS INSPECTED	Struct./Mech.	Coating
106.		
Port topside tank no. 4 (internal inspection)		
107.		
Stbd topside tank no. 4 (internal inspection)		
108.		

REMARKS:		

SURVEY OF CARGO HOLDS

ITEMS INSPECTED	Struct./Mech.	Coating
109.		
Cargo Hold no. 1 – Port – Topside tank, hopper tank, port side shell plating and frames		
110.		
Cargo Hold no. 1 – For'd blkhd corrugations, lower stool, upper stool.		
111.		
Cargo Hold no. 1 – Stbd – Topside tank, hopper tank, port side shell plating and frames		
REMARKS:		

ITEMS INSPECTED		Struct./Mech.	Coating
112.			
Cargo Hold no. 1 – After blkhd corrugations, lower stool, upper stool.			
113.			
Cargo Hold no. 1 – Tanktop, incl. bilge suctions (P&S)			
114.			

REMARKS:			

ITEMS INSPECTED		Struct./Mech.	Coating
115.			
Cargo Hold no. 2 – Port – Topside tank, hopper tank plating, port side shell plating and frames			
116.			
Cargo Hold no. 2 – For'd blkhd corrugations, lower stool, upper stool.			
117.			
Cargo Hold no. 2 – Stbd – Topside tank, hopper tank plating, port side shell plating and frames			
REMARKS:			

ITEMS INSPECTED		Struct./Mech.	Coating
118.			
Cargo Hold no. 2 – After blkhd corrugations, lower stool, upper stool.			
119.			
Cargo Hold no. 2 – Tanktop, incl. bilge suctions (P&S)			
120.			

REMARKS:			

ITEMS INSPECTED	Struct./Mech.	Coating
121.		
Cargo Hold no. 3 – Port – Topside tank, hopper tank plating, port side shell plating and frames		
122.		
Cargo Hold no. 3 – For'd blkhd corrugations, lower stool, upper stool.		
123.		
Cargo Hold no. 3 – Stbd – Topside tank, hopper tank plating, port side shell plating and frames		
REMARKS:		

ITEMS INSPECTED	Struct./Mech.	Coating
124.		
Cargo Hold no. 3 – After blkhd corrugations, lower stool, upper stool.		
125.		
Cargo Hold no. 3 – Tanktop, incl. bilge surctions (P&S)		
126.		

REMARKS:		

ITEMS INSPECTED	Struct./Mech.	Coating
127.		
Cargo Hold no. 4 – Port – Topside tank, hopper tank plating, port side shell plating and frames		
128.		
Cargo Hold no. 4 – For'd blkhd corrugations, lower stool, upper stool.		
129.		
Cargo Hold no. 4 – Stbd – Topside tank, hopper tank plating, port side shell plating and frames		
REMARKS:		

ITEMS INSPECTED	Struct./Mech.	Coating
130.		
Cargo Hold no. 4 – After blkhd corrugations, lower stool, upper stool.		
131.		
Cargo Hold no. 4 – Tanktop, incl. bilge suctions (P&S)		
132.		

REMARKS:		

SURVEY OF HOPPER TANKS: PORT (INTERNAL INSPECTION)

ITEMS INSPECTED	Struct./Mech.	Coating
133.		
Port hopper tank no. 1 (internal inspection)		
134.		
Stbd hopper tank no. 1 (internal inspection)		
135.		

REMARKS:		

ITEMS INSPECTED	Struct./Mech.	Coating
136.		
Port hopper tank no. 2 (internal inspection)		
137.		
Stbd hopper tank no. 2 (internal inspection)		
138.		

REMARKS:		

ITEMS INSPECTED	Struct./Mech.	Coating
139.		
Port hopper tank no. 3 (internal inspection)		
140.		
Stbd hopper tank no. 3 (internal inspection)		
141.		

REMARKS:		

ITEMS INSPECTED	Struct./Mech.	Coating
142.		
Port hopper tank no. 4 (internal inspection)		
143.		
Stbd hopper tank no. 4 (internal inspection)		
144.		

REMARKS:		

SURVEY OF DOUBLE BOTTOM TANKS: PORT (INTERNAL INSPECTION)

ITEMS INSPECTED	Struct./Mech.	Coating
145.		
Cargo Hold no. 1 – Port side D.B. tanks (internal inspection)		
146.		
Cargo Hold no. 1 – Stbd side D.B. tanks (internal inspection)		
147.		
Cargo Hold no. 1 – Center D.B. tanks (internal inspection)		
REMARKS:		

ITEMS INSPECTED	Struct./Mech.	Coating
148.		
Cargo Hold no. 2 – Port side D.B. tanks (internal inspection)		
149.		
Cargo Hold no. 2 – Stbd side D.B. tanks (internal inspection)		
150.		
Cargo Hold no. 2 – Center D.B. tanks (internal inspection)		
REMARKS:		

ITEMS INSPECTED	Struct./Mech.	Coating
151.		
Cargo Hold no. 3 – Port side D.B. tanks (internal inspection)		
152.		
Cargo Hold no. 3 – Stbd side D.B. tanks (internal inspection)		
153.		
Cargo Hold no. 3 – Center D.B. tanks (internal inspection)		
REMARKS:		

ITEMS INSPECTED	Struct./Mech.	Coating
154.		
Cargo Hold no. 4 – Port side D.B. tanks (internal inspection)		
155.		
Cargo Hold no. 4 – Stbd side D.B. tanks (internal inspection)		
156.		
Cargo Hold no. 4 – Center D.B. tanks (internal inspection)		
REMARKS:		

ITEMS INSPECTED	Struct./Mech.	Coating
157.		
Duct keel (internal inspection)		
158.		
After Peak tank (internal inspection)		
159.		

REMARKS:		

SURVEY OF CARGO HOLDS FIRE EXTINGUISHING SYSTEMS

ITEMS INSPECTED	Struct./Mech.	Coating
160.		
Cargo Hold no. 1 – Fixed fire extinguishing systems		
161.		
Cargo Hold no. 1 – Fire detection systems		
162.		

REMARKS:		

ITEMS INSPECTED	Struct./Mech.	Coating
163.		
Cargo Hold no. 2 – Fixed fire extinguishing systems		
164.		
Cargo Hold no. 2 – Fire detection systems		
165.		

REMARKS:		

ITEMS INSPECTED	Struct./Mech.	Coating
166.		
Cargo Hold no. 3 – Fixed fire extinguishing systems		
167.		
Cargo Hold no. 3 – Fire detection systems		
168.		

REMARKS:		

ITEMS INSPECTED	Struct./Mech.	Coating
169.		
Cargo Hold no. 4 – Fixed fire extinguishing systems		
170.		
Cargo Hold no. 4 – Fire detection systems		
171.		

REMARKS:		

SURVEY OF DOUBLE SKIN SIDE TANKS (INTERNAL)

ITEMS INSPECTED	Struct./Mech.	Coating
172.		
Cargo Hold no. 1 – Port side Double skin tank (internal inspection)		
173.		
Cargo Hold no. 1 – Stbd side Double skin tank (internal inspection)		
174.		

REMARKS:		

ITEMS INSPECTED	Struct./Mech.	Coating
175.		
Cargo Hold no. 2 – Port side Double skin tank (internal inspection)		
176.		
Cargo Hold no. 2 – Stbd side Double skin tank (internal inspection)		
177.		

REMARKS:		

ITEMS INSPECTED	Struct./Mech.	Coating
178.		
Cargo Hold no. 3 – Port side Double skin tank (internal inspection)		
179.		
Cargo Hold no. 3 – Stbd side Double skin tank (internal inspection)		
180.		

REMARKS:		

ITEMS INSPECTED	Struct./Mech.	Coating
181.		
Cargo Hold no. 4 – Port side Double skin tank (internal inspection)		
182.		
Cargo Hold no. 4 – Stbd side Double skin tank (internal inspection)		
183.		

REMARKS:		

SURVEY OF UPPER AND LOWER STOOLS (INTERNAL INSPECTION)

ITEMS INSPECTED	Struct./Mech.	Coating
184.		
Cargo Hold no. 1 – Upper stool (internal inspection)		
185.		
Cargo Hold no. 1 – Lower stool (internal inspection)		
186.		

REMARKS:		

ITEMS INSPECTED	Struct./Mech.	Coating
187.		
Cargo Hold no. 2 – Upper stool (internal inspection)		
188.		
Cargo Hold no. 2 – Lower stool (internal inspection)		
189.		

REMARKS:		

ITEMS INSPECTED	Struct./Mech.	Coating
190.		
Cargo Hold no. 3 – Upper stool (internal inspection)		
191.		
Cargo Hold no. 3 – Lower stool (internal inspection)		
192.		

REMARKS:		

ITEMS INSPECTED	Struct./Mech.	Coating
193.		
Cargo Hold no. 4 – Upper stool (internal inspection)		
194.		
Cargo Hold no. 4 – Lower stool (internal inspection)		
195.		

REMARKS:		

SURVEY OF ACCOMMODATION

ITEMS INSPECTED	Struct./Mech.	Coating
196.		
Crew accommodation		
197.		
Officers' accommodation		
198.		
Accommodation lifts		
REMARKS:		

ITEMS INSPECTED	Struct./Mech.	Coating
199.		
Passenger accommodation		
200.		
Galley		
201.		
Accommodation (external condition incl. ladders)		
REMARKS:		

ITEMS INSPECTED	Struct./Mech.	Coating
202.		
HVAC (Heating, Ventilating, Air Conditioning)		
203.		
Accommodation, internal staircases and stairs		
204.		

REMARKS:		

ITEMS INSPECTED	Struct./Mech.	Coating
205.		
Refrigerated store rooms (Dairy)		
206.		
Refrigerated store rooms (Fish)		
207.		
Refrigerated store rooms (Meat)		
REMARKS:		

ITEMS INSPECTED	Struct./Mech.	Coating
208.		
Refrigerated store rooms (Defrosting)		
209.		
Refrigerated store rooms (Vegetables)		
210.		
Refrigerated store rooms (Dry)		
REMARKS:		

ITEMS INSPECTED	Struct./Mech.	Coating
211.		
Portable safety appliances		
212.		
Fixed fire extinguishing systems		
213.		

REMARKS:		

SURVEY OF DECK OUTFIT

ITEMS INSPECTED	Struct./Mech.	Coating
214		
Bow doors		
215		
Stern doors		
216		
Side doors		
REMARKS:		

ITEMS INSPECTED	Struct./Mech.	Coating
217		
Side doors		
218		
Funnel		
219		

REMARKS:		

SURVEY OF LIFE SAVING EQUIPMENT

ITEMS INSPECTED	Struct./Mech.	Coating
220		
Lifeboats		
221		
Davits		
222		
Rescue boat		
REMARKS:		

ITEMS INSPECTED	Struct./Mech.	Coating
223		
Lifebuoys		
224		
Liferafts		
225		

REMARKS:		

SURVEY OF MAIN MACHINERY

ITEMS INSPECTED	Struct./Mech.	Coating
226.		
Main Engine (Port)		
227.		
Main Engine – Cylinder Liner wear		
228.		
Main Engine – Crankshaft deflection		
REMARKS:		

ITEMS INSPECTED	Struct./Mech.	Coating
229.		
Main Engine (Stbd)		
230.		
Main Engine – Cylinder Liner wear		
231.		
Main Engine – Crankshaft deflection		
REMARKS:		

ITEMS INSPECTED	Struct./Mech.	Coating
232.		
Main Engine (Port & Stbd) – Spare parts		
233.		
Main Engine (Port) – Luboil consumption		
234.		
Main Engine (Stbd) – Luboil consumption		
REMARKS:		

ITEMS INSPECTED	Struct./Mech.	Coating
235.		
Boiler (Port)		
236.		
Boiler (Stbd)		
237.		

REMARKS:		

ITEMS INSPECTED	Struct./Mech.	Coating
238.		
Reduction Gear ad Clutches		
239.		

240.		

REMARKS:		

ITEMS INSPECTED	Struct./Mech.	Coating
241.		
Piping		
242.		
Shafting and thrusters – port M.E.		
243.		
Shafting and thrusters – stbd M.E.		
REMARKS:		

SURVEY OF AUXILIARY DIESEL GENERATORS

ITEMS INSPECTED	Struct./Mech.	Coating
244.		
Auxiliary Diesel engine (no. 1)		
245.		
Auxiliary Diesel engine (no. 1) – Liner wear		
246.		
Auxiliary Diesel engine (no. 1) – Crankshaft deflections, spare parts and luboil consumption		
REMARKS:		

ITEMS INSPECTED	Struct./Mech.	Coating
247.		
Auxiliary Diesel engine (no. 2)		
248.		
Auxiliary Diesel engine (no. 2) – Liner wear		
249.		
Auxiliary Diesel engine (no. 2) – Crankshaft deflections, spare parts and luboil consumption		
REMARKS:		

ITEMS INSPECTED	Struct./Mech.	Coating
250.		
Auxiliary Diesel engine (no. 3)		
251.		
Auxiliary Diesel engine (no. 3) – Liner wear		
252.		
Auxiliary Diesel engine (no. 3) – Crankshaft deflections, spare parts and luboil consumption		
REMARKS:		

SURVEY OF OTHER AUXILIARY EQUIPMENT

ITEMS INSPECTED	Struct./Mech.	Coating
253.		
Condensers		
254.		
Evaporators		
255.		
Compressors		
REMARKS:		

ITEMS INSPECTED	Struct./Mech.	Coating
256.		
Fresh water generator		
257.		
Incinerator		
258.		
Pumps		
REMARKS:		

ITEMS INSPECTED	Struct./Mech.	Coating
259.		
Tailshaft(s) (last drydocking, condition and clearances)		
260.		
Tailshaft(s) (spare shafts)		
261.		
Propeller(s) (last drydocking, condition and clearances)		
REMARKS:		

ITEMS INSPECTED	Struct./Mech.	Coating
262.		
Steering gear and repeaters		
263.		
Communication systems with navigation bridge		
264.		
Emergency steering gear		
REMARKS:		

ITEMS INSPECTED	Struct./Mech.	Coating
265.		
Rudder(s) – (last drydocking, condition and clearances)		
266.		

267.		

REMARKS:		

SURVEY OF SAFETY EQUIPMENT

ITEMS INSPECTED	Struct./Mech.	Coating
268.		
Fire pump		
269.		
Emergency fire pump		
270.		

REMARKS:		

ITEMS INSPECTED	Struct./Mech.	Coating
271.		
Portable fire extinguishers (accommodation)		
272.		
Portable fire extinguishers (engine-room)		
273.		
Portable fire extinguishers (deck)		
REMARKS:		

ITEMS INSPECTED	Struct./Mech.	Coating
274.		
Fixed fire extinguishers (accommodation)		
275.		
Fixed fire extinguishers (engine-room)		
276.		
Fixed fire extinguishers (deck)		
REMARKS:		

ITEMS INSPECTED	Struct./Mech.	Coating
277.		
Fixed extinguishing systems (cargo holds)		
278.		
Fixed extinguishers (paint-room)		
279.		
Fixed extinguishers (CO2)		
REMARKS:		

ITEMS INSPECTED	Struct./Mech.	Coating
280.		
Fire detection systems (accommodation)		
281.		
Fire detection systems (engine-room)		
282.		
Fire detection systems (cargo holds)		
REMARKS:		

ITEMS INSPECTED	Struct./Mech.	Coating
283.		
Fuel shut-off devices (engine-room)		
284.		
Fuel shut-off devices (emergency generator)		
285.		
Fuel shut-off devices (other services)		
REMARKS:		

SURVEY OF ELECTRICAL SYSTEMS

ITEMS INSPECTED	Struct./Mech.	Coating
286.		
Instrumentation		
287.		
Controls		
288.		
Internal cleanliness		
REMARKS:		

ITEMS INSPECTED	Struct./Mech.	Coating
289.		
Wiring and cable termination		
290.		
Names and warning plates		
291.		
Emergency switchboard and instrumentation		
REMARKS:		

ITEMS INSPECTED	Struct./Mech.	Coating
292.		
Emergency switchboard, controls		
293.		
Emergency switchboard, wire and cable termination		
294.		
Emergency switchboard, automatic emergency generator starting		
REMARKS:		

ITEMS INSPECTED	Struct./Mech.	Coating
295.		
Shore connection		
296.		
Distribution boards – Starters, Control panels, Consoles		
297.		

REMARKS:		

ITEMS INSPECTED	Struct./Mech.	Coating
298.		
Lighting – Accommodation external		
299.		
Lighting – Accommodation internal		
300.		
Lighting – Weather Decks		
REMARKS:		

ITEMS INSPECTED	Struct./Mech.	Coating
301.		
Motors – Cables, glands, cleanliness, vibrations		
302.		
Lighting – Accommodation Engine Room		
303.		

REMARKS:		

SURVEY OF OTHER EQUIPMENT

ITEMS INSPECTED	Struct./Mech.	Coating
304.		
HF/M Radio		
305.		
SATCOM		
306.		
VHF		
REMARKS:		

ITEMS INSPECTED	Struct./Mech.	Coating
307.		
Lifeboat HF set(s)		
308.		
EPIRBS/SARTS		
309.		
Weatherfax		
REMARKS:		

SURVEY OF OTHER EQUIPMENT: INTERNAL COMMUNICATIONS

ITEMS INSPECTED	Struct./Mech.	Coating
310.		
Autotelephones		
311.		
Battery/Sound powered telephones		
312.		

REMARKS:		

ITEMS INSPECTED	Struct./Mech.	Coating
313.		
PA system		
314.		
Walky Talkies		
315.		

REMARKS:		

SURVEY OF OTHER EQUIPMENT: NAVIGATIONAL CHARTS

ITEMS INSPECTED	Struct./Mech.	Coating
316.		
Sufficiency for current trade		
317.		
Charts last updated		
318.		
Chart corrections onboard		
REMARKS:		

SURVEY OF OTHER EQUIPMENT: NAVIGATION INSTRUMENTS

ITEMS INSPECTED	Struct./Mech.	Coating
319.		
Autopilot		
320.		
Chronometers		
321.		
Anemometer		
REMARKS:		

ITEMS INSPECTED	Struct./Mech.	Coating
322.		
Echo Sounder(s)		
323.		
Gyro compass		
324.		
Loading computer		
REMARKS:		

ITEMS INSPECTED	Struct./Mech.	Coating
325.		
Magnetic compass		
326.		
Navigation and signal lights		
327.		
Positioning equipment		
REMARKS:		

ITEMS INSPECTED	Struct./Mech.	Coating
328.		
Radars		
329.		
Rudder angle indicator		
330.		

REMARKS:		

OTHER INSPECTIONS: LOG-BOOKS

ITEMS INSPECTED	Struct./Mech.	Coating
331.		
Log-Books (Deck)		
332.		
Log-Books (Engine)		
333.		

REMARKS:		

OTHER INSPECTIONS: PORT STATE CONTROL INSPECTIONS

ITEMS INSPECTED	Struct./Mech.	Coating
334.		
Port State Control		
335.		

336.		

REMARKS:		

OTHER INSPECTIONS: SHIP SAFETY MANAGEMENT

ITEMS INSPECTED	Struct./Mech.	Coating
337.		
Ship Safety Management		
338.		

339.		

REMARKS:		

OTHER INSPECTIONS: SHIP THICKNESS MEASUREMENTS

ITEMS INSPECTED	Struct./Mech.	Coating
340.		
Records of thickness measurements		
341.		

342.		

REMARKS:		

OTHER INSPECTIONS: SHIP CLASS STATUS

Class Survey	Assigned Date	Next Due Date	R&CC
1. Annual Survey			No
			Yes
2. Special Survey			No
			Yes
3. Docking Survey			No
			Yes
4. Boiler Survey			No
			Yes
5. Machinery Survey			No
			Yes
6. Steampipe Survey			No
			Yes
7. Tailshaft Survey			No
			Yes

8.			No
			Yes
9.			No
			Yes

R&CC: RECOMMENDATIONS AND CONDITIONS OF CLASS
OTHER INSPECTIONS: SHIP TRADING AND OTHER CERTIFICATES

CERTIFICATE/ DOCUMENT	Issued on	Expires on:	Endorsed on:
Issuing Authority:			
1. Certificate of Registry			
Issuing Authority:			
2. International Tonnage			
Issuing Authority:			
3. Panama Canal Tonnage			
Issuing Authority:			
4. Suez Canal Tonnage			
Issuing Authority:			
5. International Load Line			
Issuing Authority:			
6. Safety Construction			
Issuing Authority:			
7. Safety Equipment			
Issuing Authority:			
8. GMDSS			
Issuing Authority:			
9. ISM CODE			
Issuing Authority:			
10. Minimum Manning Certificate			
Issuing Authority:			
11. IOPP Certificate			
Issuing Authority:			
12. Oil Pollution Cert. of Insurance			
Issuing Authority:			

13. Deratting			
Issuing Authority:			
14. Carriage of Grain			
Issuing Authority:			
15. IMO Certificate of Fitness			
Issuing Authority:			
16. Noxious Liquid Substances			
Issuing Authority:			
17. Carriage of Dangerous Goods			
Issuing Authority:			
18. Port State Control			
Issuing Authority:			
19. Class Hull Survey			
Issuing Authority:			
20. Class Machinery Survey			
Issuing Authority:			
21. U.S.C.G. Letter of Compliance			
Issuing Authority:			
22. U.S.C.G. Cert. of Financial Responsibility			

LIST OF DEFECTS FOUND UPON INSPECTION (PART THREE)

LIST OF DEFECTS FOUND UPON INSPECTION

SHIP'S NAME	
IMO/LR NUMBER	
PORT OF SURVEY	
DATE OF SURVEY	

The signature of the Master, below, confirms that he has received a copy of this document

Defect Serial Number	1	Number in Part 2 of the report	

Defect Serial Number	2	Number in Part 2 of the report	

Defect Serial Number	3	Number in Part 2 of the report	

Defect Serial Number	4	Number in Part 2 of the report	

Defect Serial Number	5	Number in Part 2 of the report	

Defect Serial Number	6	Number in Part 2 of the report	

Defect Serial Number	7	Number in Part 2 of the report	

Defect Serial Number	8	Number in Part 2 of the report	

Signature of Ship's Master
and Ship's Stamp

Signature of Attending Surveyor

BUNKERS' SURVEY (PART FOUR)

The undersigned surveyor attended onboard the:

Motor-vessel

Whilst she was lying alongside berth no:

In the port of

On the

[005.01:Adjust the wording to reflect actual port berh, port name and date/time as mm/dd/yyyy, time as hh:mm]

On boarding, we met with the Master, the Chief-Officer and her Chief-Engineer, to whom we explained the purpose of our attendance.

To obtain the soundings of the various tanks free from any effect due to trim changes (such changes could come about at the time of cargo operations) we elected to carry out this operation as the first part of our inspection, prior to commencement of any cargo loading, and bunkering operations.

Accompanied by the ship's Chief-Engineer we witnessed the sounding of all tanks presented to us as carrying either Fuel Oil or Diesel Oil.

Once the soundings of the various fuels-tanks had been obtained we utilized the ship's calibration tables, the fuel densities shown in the last supply receipts and ASTM Tables "D-1250 / 80".

The results of our calculations may be found in the attached tables.

The quantities R.O.B. on/at: (date)   (hours)

FUEL OIL:	Metric tons
DIESEL OIL:	Metric tons

EXPLANATORY NOTES ON PART NOS. 1, 2, 3, AND 4
PART 1: SURVEY REPORT PREAMBLE

The wording of the various report sections is self-explanatory.

PART 2: SHIP CONDITION SURVEY

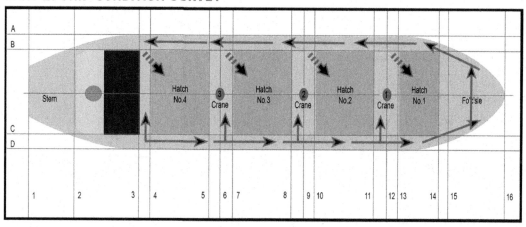

Deck inspection, suggested rotation.

The numbers, above, indicate the rotation of the areas to be inspected on the weather-deck, stbd-side. A similar rotation is repeated on the weather-deck, port-side.

The sections of the condition survey are color coded as under:

Color Code

HULL

MACHINERY

OTHER EQUIPMENT

The suggested condition-notations regarding the ship's structural and/or mechanical parts are as per the following:

Excellent	Good	Average	Substandard
	Minor wear and tear. May be left as is.	Wear and Tear Evident. Corrective action an Option.	Repairs/Renewals required now. Entered in defects' list

The suggested condition-notations regarding the coating of spaces and/or tanks may be found below:

EXCELLENT	GOOD	AVERAGE	POOR
	MINOR SPOT RUST ON NO MORE THAN 20% OF AREA CONSIDERED	LIGHT RUST ON 20% OR MORE OF AREA CONSIDERED	COATING BROKEN DOWN ON 20% OR MORE OF AREA CONSIDERED WITH HARD SCALE OVER 10% OR MORE OF THAT AREA

Each section of the survey report contains a table, as the one shown below. This table covers the condition of each part listed with regard to:

(1) the structural/mechanical condition and
(2) the coating condition of the item considered.

In addition, each of these tables provides a section headed "Remarks," which serves to record any notes the surveyor considers to be appropriate.

ITEMS INSPECTED	Struct./Mech.	Coating
343.		
Port side bow shell plating		
344.		
Port side shell plating		
345.		
Port side stern shell plating		
REMARKS:		

Should the condition of an item be so poor that it warrants a defect notation, the surveyor ought to ensure that the particular item is recorded in the separate list of defects, in "Part 3."

The drop down list for the Coating Condition rating works in a manner similar to that of the first drop down list.

The "Ship Condition Survey" concludes with a record of the ship's certificates. The surveyor must scrutinize these documents and enter any Recommendations, or Conditions of Class that may be currently outstanding.

PART 3: SHIP DEFECTS' LIST

A sample of the defects list follows below (The first defect has been completed to clarify the way this section of the report is used. Note that the number corresponding to this item, in Part 2, has also been logged):

SHIP DEFECTS' LIST

SHIP'S NAME	
IMO/LR NUMBER	
PORT OF SURVEY	
DATE OF SURVEY	

The signature of the Master, below, confirms that he has received a copy of this document

Defect Serial Number	**1**	Number in Part 2 of the report	**78**
The crossdeck between cargo-holds no. 2 and 3 was found holed in numerous areas, with underside stiffeners exposed and subject to heavy wear. In consideration of these findings, it is recommended that permanent repairs, to class satisfaction, should be carried out before the ship's departure from this port.			

Signature of Ship's Master
And Ship's Stamp

Signature of Attending Surveyor

PART 4: SHIP BUNKERS' SURVEY

The wording of the bunkers' survey report is self-explanatory.

BUNKER FUEL MEASUREMENT REPORT

Petroleum Measurement Tables utilized: ASTM D 1250-80 Table 54B and volume XII, Table 21

Ship's Name:		**ALIBION 35**	
Port of Survey:		PORT RASHID, DUBAI, U.A.E.	

Survey Date:	27.01.2012	Time: Start:	0830 hours	Finish:	1100 hours

Draft Aft:	6.00 m	Draft Amidships, Port:	5.20
Draft For'd:	5.00 m	Draft Amidships, Stbd:	5.20
Trim:	1.00 m	Heel:	Nil

The receipt furnished by the ship's Chief-Engineer for this fuel stated the following:

API Gravity at 60 F	Relative Density 60/60 F	Density at 15 C
-----	-----	0.9706 and 0.9780
Sea Condition:	Calm	

Product surveyed	FUEL OL

Tank Designation	Sounding Obs.	Sounding Corr.	Temp Obs.	Density 15/15C	Volume Trim Corr.	Density In Air	Volume Corr. Factor	Product Weight In Air
TANK 23 (S)	7.59		36	0.9706	246.00	0.9695	0.9852	234.970
TANK 27 (S)	7.93		36	0.9780	297.31	0.9769	0.9854	286.201
TANK 28 (P)	8.04		36	0.9780	303.49	0.9769	0.9854	292.149
TANK 24 (P)	7.89		36	0.9706	258.81	0.9695	0.9852	247.206
TANK 20 (P)	0.07		36	0.9706	1.07	0.9695	0.9852	1.021
TANK 26 (P)	6.73		36	0.9780	188.80	0.9769	0.9854	181.750
TANK 25 (S)	6.60		36	0.9780	166.92	0.9769	0.9854	160.683
TANK 52			75	0.9706	15.00	0.9695	0.9576	13.926
TANK 48			45	0.9706	48.00	0.9695	0.9789	45.554
TANK 47			42	0.9706	30.00	0.9695	0.9810	28.532
TOTAL FUEL OIL IN ALL TANKS SOUNDED, METRIC TONS, IN AIR								**1,491.992**

The receipt furnished by the ship's Chief-Engineer for this fuel stated the following:

API Gravity at 60 F	Relative Density 60/60 F	Density at 15 C
-----	-----	0.8704

Product surveyed	DIESEL OL

Tank Designation	Sounding Obs.	Sounding Corr.	Temp. Obs.	Density 15/15C	Volume Trim Corr.	Density In Air	Volume Corr. Factor	Product Weight In Air
TANK PORT	1.93		28	0.8704	33.05	0.8693	0.9895	28.429
TANK STBD	4.04		28	0.8704	50.06	0.8693	0.9895	43.060
TANK SERVICE	2.53		26	0.8704	5.43	0.8693	0.9911	4.678
TOTAL DIESEL OIL IN ALL TANKS SOUNDED, METRIC TONS, IN AIR								**76.167**

List of References

http://marine.man.eu/docs/librariesprovider6/technical-papers/propulsion-of-vlcc.pdf?sfvrsn=22

http://www.iso.org/iso/home/standards/management-standards.htm

http://www.imo.org/Publications/SupplementsAndCDs/Documents/Certificatesonboardships.pdf

http://www.iacs.org.uk/

INTERNATIONAL ASSOCIATION OF CLASSIFICATION SOCIETIES LTD.

Safer and Cleaner Shipping

261

https://www.google.co.uk/webhp?sourceid=chrome-instant&ion=1&espv=2&ie=UTF-8#q=DATAMONITOR%2C+UK+Marine+Insurance+2012

UK Marine Insurance 2012

http://enraf.ru/userfiles/File/4416650_rev4.pdf

http://www.mpa.gov.sg/sites/port_and_shipping/port/bunkering/bunkering_standards/singapore_standard_code_of_practice_for_bunkering-ss_60.page

https://www.bimco.org/~/media/Products/Manuals-Pamphlets/Bunkering_Guide/Bunkering_Pamphlet.ashx

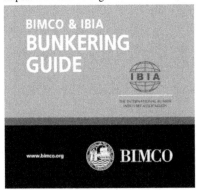

ITIC—GUIDELINES ON INCORPORATING STANDARD TERMS AND CONDITIONS
http://www.itic-insure.com/rules-publications/article/guidelines-on-incorporating-standard-terms-and-conditions-129819/

TUGS, TOWBOATS, AND TOWING
Edward M. Brady

MARINE SALVAGE OPERATIONS
Edward M. Brady

TANKER OPERATIONS—A HANDBOOK FOR THE SHIP'S OFFICER
G. S. Marton

PRINCIPLES OF MARITIME LAW
Hodges and Hill

STEEL—CARRIAGE BY SEA
Sparks

THE THEORY AND PRACTICE OF SEAMANSHIP
Graham Danton

THE DESIGN OF MARINE SCREW PROPELLERS
T. P. O'Brien

MATERIALS FOR MARINE MACHINERY
Frederick and Capper

STEEL PLATES
Bethlehem Steel Corporation, Handbook 466-D

STRUCTURAL SHAPES
Bethlehem Steel Corporation, Catalog 2747

REFRIGERATION AT SEA
Munton, Stott

SHIPPING LAW
N. J. J. Gaskell, C. Debattista, and R. J. Swatton

MARINE UNDERWRITING
The Chartered Insurance Institute

MODERN SHIP DESIGN
Thomas C. Gillmer

COATINGS AND INSPECTION MANUAL
Jotun

ABS—GRADING BOOKLET HIMP
American Bureau of Shipping

CLASS NK—GOOD MAINTENANCE ON BOARD SHIPS
Nippon Kaiji Kyokai

CLASS NK—PORT STATE INSPECTIONS POCKET CHECKLIST
Nippon Kaiji Kyokai

LR—ESP GUIDANCE BOOKLET FOR ALL SHIPS
Lloyd's Register

INSPECTION GRADING CRITERIA FOR THE ABS HULL INSPECTION AND MAINTENANCE PROGRAM (HIMP)
American Bureau of Shipping

ONBOARD ROUTINE MAINTENANCE CHECK SHEET
American Bureau of Shipping

HULL STRUCTURAL DESIGN, SHIPS WITH LENGTH 100 METERS AND ABOVE
Det Norske Veritas

HULL STRUCTURAL DESIGN, SHIPS WITH LENGTH 100 METERS AND ABOVE
Det Norske Veritas

ENHANCED SURVEY PROGRAM (ESP) GUIDANCE BOOKLET FOR ALL SHIP TYPES IN PREPARATION FOR SPECIAL SURVEY
Lloyd's Register

GUIDANCE INFORMATION ON SPARE GEAR
Lloyd's Register

**RULES FOR CLASSIFICATION AND CONSTRUCTION OF SHIPS
SHIP TECHNOLOGY**
Germanischer Lloyd

GUIDANCE RELATING TO THE RULES FOR THE CLASSIFICATION OF STEEL SHIPS PART 3—HULL STRUCTURES
Korean Register

BULK CARRIERS AND OIL TANKERS, TECHNICAL BACKGROUND RULE REFERENCE
IACS

WHAT ARE CLASSIFICATION SOCIETIES
IACS

CLASSIFICATION SOCIETIES—WHAT, WHY, AND HOW
IACS

REQUIREMENTS CONCERNING MOORING, ANCHORING, AND TOWING
IACS

REQUIREMENTS CONCERNING SURVEY AND CERTIFICATION
IACS

REVISED LIST OF CERTIFICATES AND DOCUMENTS REQUIRED TO BE CARRIED ON BOARD SHIPS
IMO

CHARTERING TERMS
Project Professional Group

Index

Note: Page numbers followed by *f* indicate figures and *t* indicate tables.